Ana Rodríguez Bernaldo de Quirós, Antía Lestido Cardama, Raquel Sendón,
Verónica García Ibarra
Food Contamination by Packaging

Also of interest

Metals in Wastes.
Wieszczycka, Tylkowski, Staszak, 2018
ISBN 978-3-11-054628-6, e-ISBN 978-3-11-054706-1

Inorganic Trace Analytics.
Trace Element Analysis and Speciation
Matusiewicz, Bulska, 2017
ISBN 978-3-11-037194-9, e-ISBN 978-3-11-036673-0

Organic Trace Analysis.
Niessner, Schaeffer, 2017
ISBN 978-3-11-044114-7, e-ISBN 978-3-11-044115-4

Elemental Analysis.
An Introduction to Modern Spectrometric Techniques
Schlemmer, Balcaen, Todolí, Hinds, 2019
ISBN 978-3-11-050107-0, e-ISBN 978-3-11-050108-7

Ana Rodríguez Bernaldo de Quirós,
Antía Lestido Cardama, Raquel Sendón,
Verónica García Ibarra

Food Contamination by Packaging

Migration of Chemicals from Food Contact Materials

DE GRUYTER

Authors

Prof. Dr. Ana Rodríguez Bernaldo de Quirós
University of Santiago de Compostela
Dept. of Analytical Chemistry
Nutrition and Food Science
Pharmacy Faculty
15782 Santiago de Compostela
Spain
ana.rodriguez.bernaldo@usc.es

Prof. Dr. Raquel Sendón
University of Santiago de Compostela
Dept. of Analytical Chemistry
Nutrition and Food Science
Pharmacy Faculty
15782 Santiago de Compostela
Spain
raquel.sendon@usc.es

Dr. Antía Lestido Cardama
University of Santiago de Compostela
Dept. of Analytical Chemistry
Nutrition and Food Science
Pharmacy Faculty
15782 Santiago de Compostela
Spain
antia.lestido@usc.es

Dr. Verónica García Ibarra
University of Santiago de Compostela
Dept. of Analytical Chemistry
Nutrition and Food Science
Pharmacy Faculty
15782 Santiago de Compostela
Spain
va.garcia@uta.edu.ec

ISBN 978-3-11-064487-6
e-ISBN (PDF) 978-3-11-064806-5
e-ISBN (EPUB) 978-3-11-064502-6

Library of Congress Control Number: 2019936773

Bibliographic information published by the Deutsche Nationalbibliothek
The Deutsche Nationalbibliothek lists this publication in the Deutsche Nationalbibliografie; detailed bibliographic data are available on the Internet at http://dnb.dnb.de.

Typesetting: Integra Software Services Pvt. Ltd.
Printing and binding: CPI books GmbH, Leck
Cover image: Symbol of the European Union to label materials suitable for food contact, after EU Regulation (EC) No. 1935/2004. Cover designed by De Gruyter.

www.degruyter.com

Preface

The migration of substances from packaging to food is one of the main concerns associated with food contact materials (FCM) for food safety authorities, and packaging materials constitute a potential source of contaminants to which the consumer will be exposed to through their diet.

These contaminants comprise a wide variety of chemical compounds such as phthalates and plasticisers, bisphenol A (BPA) and photoinitiators, among others, and for some of them there is evidence of adverse effects on health.

This book provides an overview of the main packaging contaminants, and particular attention is paid to the chemical aspects and analytical methodologies for their determination.

The first part includes Chapter 1, which presents a brief introduction to chemical migration. The second part is composed of six chapters; Chapters 2–5 are devoted to BPA, melamine, phthalates, and other plasticisers and photoinitiators, respectively. Chapter 6 reviews other chemical substances that can originate from FCM, including perfluorochemicals, saturated and aromatic hydrocarbons (mineral oil saturated hydrocarbons and mineral oil aromatic hydrocarbons) from mineral oils, other bisphenol-related compounds, nanoparticles and primary aromatic amines; and Chapter 7 focuses on nonintentionally added substances. The final part includes Chapter 8 that deals with the estimation of exposure to migrants from FCM.

https://doi.org/10.1515/9783110648065-201

Acknowledgements

The work was financially supported by the "Ministerio de Economía y Competitividad" and by "Fondo Europeo de Desarrollo Regional (FEDER), Ref. No. AGL2015-69609-P "MIGRAEXPO" (MINECO/FEDER, UE).

https://doi.org/10.1515/9783110648065-202

Contents

1 Introduction

Packaging plays an essential role protecting food from chemical contamination and microbial spoilage [1–3]. However, from a food safety point of view, the main concern associated with packaging is the chemical migration of harmful substances from the packaging into the food, which could constitute a risk to consumer health [1]. Migration is a mass transfer phenomenon where the substance transferred from the packaging to the foodstuff is called the migrant. This process obeys Fick's laws of diffusion and depends upon several factors including: type of material, nature of food, physico-chemical properties of migrant, time and temperature of contact and so on [1, 4–6]. The key parameters of the migration process are the diffusion and partition coefficients. The partition coefficient ($K_{P/F}$) represents the relative solubility of the migrant between the polymer and the food once equilibrium is reached. Hence, values of $K_{P/F} > 1$ indicate that the concentration of the migrant is greater in the polymer than in the food. On the other hand, the diffusion coefficient describes the kinetics of the migration [7–9].

Special attention has been paid to low molecular weight substances since it is generally accepted that high molecular weight compounds (> 1,000 Da) are not absorbed through the gastrointestinal tract [10]. In addition, chemical migration can also negatively affect the quality of the packed food and modify the organoleptic properties [1].

The experimental determination of migration is laborious, time-consuming and expensive due mainly to the complex chemical composition of food and the low concentrations of migrants detected in the foodstuffs. In order to avoid these analytical difficulties, different tools for the theoretical prediction of migration have been developed [4].

To guarantee a high level of health protection, the safety of food contact materials (FCM) should be subject to regulatory control. In this context, the Commission of the European Communities has issued regulations concerning the materials intended to be in contact with food. The Framework Regulation (EC) 1935/2004 [11] is applied to all materials and articles intended to come into contact with food. Article 3 of this regulation establishes the general requirements:

1. "Materials and articles, including active and intelligent materials and articles, shall be manufactured in compliance with good manufacturing practice so that, under normal or foreseeable conditions of use, they do not transfer their constituents to food in quantities which could: (a) endanger human health, (b) bring about an unacceptable change in the composition of the food or (c) bring about a deterioration of the organoleptic characteristics thereof.
2. The labelling, advertising and presentation of a material or article shall not mislead consumers."

https://doi.org/10.1515/9783110648065-001

Annex I of this regulation lists the groups of materials and articles that may be covered by specific measures including: active and intelligent materials and articles, adhesives, ceramics, cork, rubbers, glass, ion-exchange resins, metals and alloys, paper and board, plastics, printing inks, regenerated cellulose, silicones, textiles, varnishes and coatings, waxes and wood.

Plastics are one of the most widely used materials by the food packaging industry. Commission Regulation (EU) No 10/2011 [12] on plastic materials and articles intended to come into contact with food includes the European positive list of monomers, additives and other starting substances authorised in the manufacture of plastic FCM, as well as their restrictions, for example, the specific migration limit, and so on.

A huge variety of substances could potentially be present in packaging materials and consequently can migrate into food. Different analytical techniques, particularly chromatography, both liquid and gas coupled to mass spectrometry, for the determination of migrants have been successfully applied. At present, the development of analytical methods for the identification of nonintentionally added substances (NIAS) constitutes one of the current challenges in food packaging research [13–15].

In this book, the main packaging contaminants including: phthalates and plasticisers, photoinitiators, Bisphenol A, perfluorochemicals, primary aromatic amines and NIAS, among others are reviewed, and special emphasis is paid to the chemical aspects and analytical methodologies for their determination.

References

[1] L. Castle in *Chemical Migration and Food Contact Materials*, Eds., K.A. Barnes, C. Richard Sinclair and D.H. Watson, CRC Press, Boca Raton, FL, USA, 2007, p.1.

[2] G.L. Robertson in *Food Packaging Principles and Practice*, CRC Press Taylor & Francis Group, Boca Raton, FL, USA, 2013.

[3] D. Sun Lee, K.L. Yam and L. Piergiovanni in *Food Packaging Science and Technology*, CRC Press Taylor & Francis Group, Boca Raton, FL, USA, 2008.

[4] O. Lau and S. Wong, *Journal of Chromatography A*, 2000, 882, 255.

[5] J. Crank in *Mathematics of Diffusion*, 2nd Edition, Oxford Science Press, Oxford, UK, 1975.

[6] I.S. Arvnitoyannis and K.V. Kotsanopoulos, *Food Bioprocess and Technology*, 2014, 7, 21.

[7] E.A. Tehrany and S. Desobry, *Food Additives & Contaminants*, 2004, 21, 1186.

[8] P. Otero-Pazos, R. Sendón, B. Blanco-Fernandez, S. Blanco-Dorado, C. Alvarez-Lorenzo, A. Concheiro, I. Angulo, P. Paseiro-Losada and A.Rodríguez-Bernaldo de Quirós, *Journal of Food Science and Technology*, 2016, 53, 6, 2817.

[9] J. Maia, A. Rodríguez-Bernaldo de Quirós, R. Sendón, J.M. Cruz, A. Seiler, R. Franz, C. Simoneau, L. Castle, M. Driffield, P. Mercea, P. Ooldring, V. Tosa and P. Paseiro, *Food Additives & Contaminants: Part A*, 2016, 33, 4, 715.

[10] *Note for Guidance for Petitioners Presenting an Application for the Safety Assessment of a Substance to be used in Food Contact Materials Prior to its Authorisation*, European Food Safety Authority, Parma, Italy, 2008. http://www.efsa.europa.eu/en/efsajournal/doc/21r.pdf

[11] Regulation (EC) No 1935/2004, of the European Parliament and of the Council of 27 October 2004 on materials and articles intended to come into contact with food and repealing directives 80/590/EEC and 89/109/EEC, Official Journal of the European Union, L, 2004, 338, 4.
[12] Commission European Regulation (EU) No 10/2011 of 14th January 2011 on Plastic Materials and Articles Intended to Come into Contact with Food, Official Journal of the European Union L, 2011, 12, 1.
[13] Y. Sanchis, V. Yusà and C. Coscollà, *Journal of Chromatography A*, 2017, 1490, 22.
[14] I.S. Arvnitoyannis and L. Bosnea, *Critical Reviews in Food Science and Nutrition*, 2004, 44, 63.
[15] J. Muncke, *Science of Total Environment*, 2009, 407, 4549.

2 Bisphenol A

2.1 Introduction

Bisphenol A (BPA) is an organic compound that consists of two phenol molecules bound by a methyl group and two methyl radicals [1].

BPA was first synthesised in 1891 by a Russian chemist Alexander P. Dianin via a condensation reaction of two phenol molecules and one acetone molecule (hence, the suffix 'A' in its name) in the presence of a catalyst such as hydrogen chloride or an ion-exchange resin. In 1930, BPA was analysed during investigations of synthetic oestrogen [2].

Nowadays, there are two methods for the manufacture of BPA: the acid-catalysed condensation of phenol and acetone using distillation technology, or with a different catalyst using purification technology [3].

BPA shows good solubility in fats, but is a moderately water-soluble compound at ambient temperature. Based on the octanol–water partition coefficient, BPA has modest capacity for bioaccumulation, which occurs only at high doses [1]. The physicochemical properties of BPA are shown in Table 2.1.

Table 2.1: Physical and chemical properties of BPA [1, 3].

Common name (acronym)	**Bisphenol A (BPA)**
CAS number	80-05-7
Molecular formula	$C_{15}H_{16}O_2$
Chemical structure	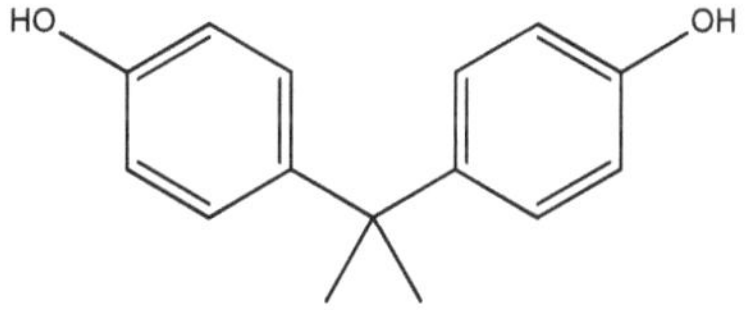
Appearance	White crystalline, granules and flakes solid
Molar weight (g/mol)	228.29
Melting point (°C)	158
Boiling point (°C)	398 (at 101.3 kPa)
Water solubility (g/m^3)	120–300 (at 20 °C)
Vapour pressure (Pa)	1.1×10^{-7}–5.3×10^{-6}

https://doi.org/10.1515/9783110648065-002

Table 2.1 (continued)

Common name (acronym)	Bisphenol A (BPA)
pK_a	9.59–11.30
Octanol–water partition coefficient $\log(K_{OW})$	2.2–3.4

2.2 Uses

Currently, BPA is one of the chemicals most commonly produced and used in industries [2, 3]. BPA production has been constantly growing due to high global demand [1]. It commonly appears in several products that are used every day by consumers, for example, in food packaging [2, 4].

Prior to the last decade, the majority of industrially produced BPA was used as a monomer or a starting substance in the manufacture of plastic materials (95%), in particular, the production of polycarbonate(s) (PC) plastics (infant feeding bottles, tableware, microwavable items, ovenware, food and drink storage containers, returnable water and milk bottles, and water supply pipes), and is the second largest outlet of epoxy resin manufacturing (used as a protective lining of the inner surface of food and beverage cans, metal lids for glass jars and bottles, and drinking water storage tanks and wine vats) [5].

PC is a thermoplastic polymer widely used due to its mechanical properties, such as transparency, low weight, easily worked, high impact resistance, low moisture absorbency, and good dimensional and thermal stability [6–8]. The production of PC is based on the reaction between the monomer of BPA and phosgene gas (carbonyl chloride), which is carried out in the presence of sodium hydroxide to form carbonate bonds in the polymer [7].

Epoxy resins are used as an internal protective liner in metal cans in order to protect the metal from corrosion and food from metal contamination to maintain the quality of canned food and beverages [2, 6].

Moreover, BPA is used as an additive for flame-retardant materials, unsaturated polyester resins, and polyacrylate, polyetherimide and polysulfone resins [5]. BPA is also employed as a stabilising material or an antioxidant in the processing of plastics such as polyvinyl chloride, commonly used in household food storage items such as transparent films; it is also found in recycled paper and paperboard used for food packaging, probably due to its use in printing inks [2, 9–11]. The use of plastic materials for food contact products is regulated; however, paper and paperboard (recycled or not) are not currently subject to regulation [10].

Recently, BPA has also been used in digital media, electrical and electronic equipment, medical tools, and dental composites and sealants [3, 9]. Moreover,

BPA has been detected in saliva from patients treated with BPA diglycidyl ether-based dental sealants [6].

2.3 Toxicity

BPA has been found in the urine of more than 90% of the US population, with higher levels in infants and children than in adults, so this compound presents a particular concern [12].

The health concern of exposure to BPA is a result of it belonging to a group of endocrine disrupting chemicals or xenoestrogens, substances that are capable of interfering with endocrine functions and disrupting the hormone balance. BPA affects the oestrogenic system and is also able to disrupt the functions of various hormones, including androgens, prolactin, leptin, insulin and thyroid hormones [2]. The oestrogenic activity of BPA was first discovered in 1993 [5]. Its chemical structure, which is similar to the natural oestrogen 17-β-oestradiol, allows it to fit into the oestrogen receptor behaving like an endocrine disrupter [13]. However, it should be noted that the affinity of BPA towards nuclear α- and β-oestrogen receptors was over 10,000- to 100,000-fold weaker in comparison to 17-β-oestradiol, so it is considered to be a weak oestrogen [5].

BPA is a noxious compound that has been related to a wide range of adverse health effects. Human exposure to BPA during various life stages has been associated with certain cancers, including mammary gland and prostate cancers; metabolic disorders such as obesity and insulin-resistant diabetes (type 2); neurobehavioural problems like autism and attention-deficit hyperactivity disorder; abnormalities in reproductive organs (a decrease in semen quality in men and earlier puberty in females) and heart diseases (heart attacks, coronary heart disease and angina) [2, 3, 8, 12, 14].

The oral intake of contaminated food appears to be the predominant route of exposure for humans, while inhalation and dermal contact may constitute the secondary exposure sources, particularly for workers employed in its synthesis [1]. Contamination with BPA may occur during production, processing, use, removal or recycling operations of end products containing BPA [15].

Contact with the skin can cause allergy symptoms and contact-dermatitis symptoms, whereas the inhalation of elevated BPA concentrations may cause altered metabolism [2].

The acute toxicity of BPA is relatively low and it is classed as a group 3 substance according to the International Agency for Research on Cancer classification, meaning it is 'not classifiable as to its carcinogenicity to humans' [1]. However, BPA has caused most concern because it exhibits oestrogenic activity even at low levels, particularly at sensitive life stages (foetuses, infants and young children) [1, 9]. The most critical period of BPA exposure is during foetal development due to low body weight and immature organ systems, which means they have an increased sensitivity to BPA and the alterations tend to be irreversible by interfering with organogenesis and histogenesis

[13, 16]. In addition, it has been found that BPA is able to cross the placenta in humans, which leads to foetal exposure to unsafe levels of BPA [12, 17]. Maternal exposure to BPA has been associated with second-generation health risks such as genital malformations, testicular abnormalities, impairment in fertility or sexual function [18].

The tolerable daily intake set by the European Union (EU) Commission and the reference dose of BPA established by the Environmental Protection Agency is 0.05 mg/kg body weight/day. This value is obtained by applying a 100-fold uncertainty factor to the nonobserved adverse effect level of 5 mg/kg [5, 19].

In 2015, the European Food Safety Authority reported in Scientific Opinion that the risk to public health, in relation to the presence of BPA in foodstuffs, was a temporary tolerably daily intake of 4 µg/kg body weight/day based on new meaningful data on toxicokinetics [15, 19].

2.4 Metabolism

In humans, orally administered BPA is rapidly absorbed from the gastrointestinal tract. An intensive biotransformation of BPA then takes place in the liver where it is conjugated with glucuronic acid via uridine diphosphate (UDP)-glucuronosyltransferase in an efficient first-pass metabolism [20]. The main protein isoform of UDP-glucuronosyltransferase responsible for BPA glucuronidation in humans is UGT2B15 [16]. The formed BPA glucuronide is rapidly delivered to the blood to reach the kidneys, where it is rapidly and completely excreted in urine (a body fluid sample appropriate for assessment of BPA exposure) within 24 h of administration [20].

The highly water-soluble BPA glucuronide is the major metabolite of BPA, which is biologically less active than free BPA due to its inability to bind to the oestrogen receptor [20]. However, BPA glucuronide becomes deconjugated by β-glucuronidase, an enzyme present in the human liver, spleen, kidneys, intestine, lung, muscles and serum, resulting in higher plasma levels of BPA in the active form [16]. The other minor metabolites of BPA can be formed by saturation of the glucuronidation pathway when higher doses of BPA are administered. In humans, studies suggest that enterohepatic circulation does not occur [20].

Due to the lipophilic character of BPA, it may accumulate in solid tissues and exert a very important effect on fat tissue, where the regulation of various metabolic processes occurs and plays an important role as an endocrine organ in humans [2].

Plants and microorganisms are also capable of transforming BPA to metabolites with lower or higher toxicity and oestrogenicity. For example, two bacterial strains, such as *Pseudomonas* sp. and *Pseudomonas putida*, exhibited the highest BPA degradation ability. It has been shown that methylated derivatives exhibited stronger toxicity, while nitro- and dinitro-derivatives exhibited lower toxicity [2].

2.5 Regulation in food contact materials

The use of BPA as a monomer or a starting substance in the manufacture of plastic materials intended to come into contact with foodstuffs is currently permitted with a specific migration limit (SML) of 0.6 mg/kg, which was set by the EU Commission [5, 11, 19]. However, in 2010, Canada was the first country to ban the importation, sale and advertising of PC baby bottles to reduce the exposure of infants and newborns to BPA [2, 14]. Thus in 2011, the European Commission (EC) restricted the use of BPA in relation to the manufacture of PC infant feeding bottles, and recommended no exposure of infants, young children, and pregnant and breastfeeding women to BPA because they are the most susceptible populations [2, 21]. This restriction was not based on any evidence of adverse effects from BPA, and was imposed only for reasons of consumer protection [15].

Recently, the EC has published a draft on the use of BPA in varnishes and coatings intended to come into contact with food, establishing an SML of 0.05 mg of BPA per kg of food but in those materials intended to come into contact with infant formula, follow-on formula, processed cereal-based food, baby food or food for special medical purposes developed to satisfy nutritional requirements and young children, its migration should not be detectable. Furthermore, this draft amended Regulation (EU) 10/2011 with regard to the use of BPA in plastic food contact materials (FCM) changing the SML from 0.6 to 0.05 mg of BPA per kg of food or not detected in plastic materials and articles specifically intended to be brought into contact with food for infants and young children [22].

In 2013, the US Food and Drug Administration ruled that BPA-based epoxy resins would no longer be used as coatings in the packaging of infant formula because its use had been abandoned by industry [15]. Since then, several components have been employed as substitutes to produce those items, for example, bisphenol S [2].

In relation to BPA legislation in developing countries, only South Africa, Brazil, Colombia and South Korea have some legislation against the use of BPA in FCM [15].

2.6 Bisphenol A in food

Food packaging allows food to be stored at different temperatures, increases its shelf life, enables safe transportation from the point of production to the point of consumption and safeguards foods from natural agents, such as air, that can alter their quality [23]. However, it should be noted that most of these materials are complex substances and are not inert [24]. Plastics are the most commonly used material in food packaging, in combination with various additives that are used to improve certain characteristics such as elasticity, flexibility, colour, resistance, durability and so on [15, 23].

Its presence in food and beverages is of particular concern because it constitutes the main route of exposure of the general population to BPA. Human exposure to BPA through the alimentary canal was estimated to be between 0.48 and 1.6 μg/kg body weight/day. BPA may be released from PC and epoxy resins used as food and beverage containers [2].

There are two possible origins for BPA to leach from PC products: migration of the residual BPA-free monomer resulting from incomplete polymerisation, which is a negligible contribution; or degradation of the polymer itself, which is the predominant mechanism responsible for BPA release from the polymer surface [6, 17, 25]. Degradation of the polymers can be accelerated at high temperatures (e.g., retaining boiling water inside or heating in the microwave), repeated use or under the acidic/basic conditions by hydrolysis of the ester bond that binds BPA molecules and increasing the d-spacing of PC [1, 3, 6, 23, 25, 26].

It was reported that dishwashing, boiling, brushing and the ageing of the bottle could lead to polymer degradation, resulting in an increase of the release of BPA. Dishwashing may increase BPA release while the detergent is in contact with the PC and the ageing of the bottle can increase BPA release due to increasing the wettability of the bottle wall [8, 25].

PC can also be degraded by aminolysis, as amines could act as a catalyst in the depolymerisation process, resulting in the release of BPA. These amines can be naturally present in almost all foods, such as fish, meat, vegetables, fruit, wine, cheese, beer and also milk [4, 25].

Several authors suggest that BPA concentrations are significantly higher in canned food compared with BPA levels in fresh food (noncanned), so eating food with limited food packaging can result in a substantial reduction in the exposure to BPA [2, 27]. Generally, the concentration of BPA is usually higher in canned fruits, vegetables, fish and meats; and lower concentrations are usually found in baby foods and liquid foods such as milk and dairy products, soft drinks and sauces [28]. However, BPA has also been found in noncanned foods as a result of BPA migration from the packaging paper and/or contamination during its production [26].

2.7 Background contamination

Laboratory contamination of BPA can occur easily as BPA can be considered nowadays as an ubiquitous chemical in the environment.

Contamination mainly arises from the materials involved in sample treatment, such as glassware or plasticware. In general, to prevent this contamination some precautionary measures are used, for example, baking glassware for 4 h at 400 °C and washing all other materials with solvents prior to use.

BPA have been found in some solvents like ultrahigh quality (Milli-Q®) water due to plastics and epoxy resins present in the water-purifying equipment. This contamination was removed by filtering the water through a hydrophobic membrane (Empore™ disk).

Solid-phase extraction (SPE) cartridges can also cause BPA contamination, probably derived from the manufacturing process or from the adhesive used to fix the needle in the syringes. These contamination sources can be effectively prevented by prewashing the cartridges with at least 15 mL of methanol or using a peristaltic pump with Viton® tubes from DuPont [5].

2.8 Sample preparation

Owing to the complexity of food samples, an extensive sample preparation is often required prior to BPA determination, including pretreatment, extraction, clean-up, concentration and sometimes derivatisation reactions [5]. Sample treatment allows enhancing the sensitivity of the method and reduces the interference of other compounds [3].

Generally, solid samples are usually first homogenised, while liquid samples are filtered and/or centrifuged. Depending on the matrix composition, other special treatments can be performed, for example, carbonated beverages have to be degassed by sonication and samples with high protein content may require its removal by precipitation. On the other hand, lipids can significantly reduce the analytical performance by affecting the active surface of the stationary phase in liquid chromatography (LC), and in gas chromatography (GC)–mass spectrometry (MS), lipids can also accumulate in the injection port, column and/or ion source. So, fatty samples require the removal of fat via several approaches, for example, liquid–liquid extraction (LLE) with *n*-heptane, trimethylpentane and *n*-hexane or by freezing the lipids in the extract followed by filtration [5].

Solvent extraction and SPE are the most widely used techniques for the isolation of BPA from solid and liquid samples, respectively, due to their simplicity and wide-ranging applicability [5].

Solvent extraction is still the most frequently employed technique for the extraction of BPA from solid foodstuffs. The selection of the most suitable solvent is based on the physicochemical properties of the analyte of interest and the characteristics of the matrix. So, acetonitrile is the solvent most commonly used for the extraction of BPA in solid foods, although other solvents like acetone, methanol and ethanol may also efficiently extract BPA. In the case of liquid foodstuffs, LLE is a technique also used, although to a lesser extent than SPE, using solvents such as ethyl acetate, chloroform and dichloromethane (DCM). Several extractions are normally carried out to achieve the complete isolation of BPA [5].

SPE is the main technique used for the extraction of BPA in liquid foodstuffs and often requires the clean-up of extracts after solvent extraction. SPE offers some advantages in comparison with LLE, for example, higher selectivity and lower solvent consumption; however, SPE is more tedious because it usually requires extra clean-up steps due to small particles present in the sample extracts that can lead to low recoveries and irreproducibility by adsorption of analytes or clogging. There are different types of sorbents for BPA extraction and its selection depends on its properties (moderately polar character and the presence of hydrogen acceptor/donor groups) and the type of food matrix:

- Nonselective sorbents: the divinylbenzene/*N*-vinylpyrrolidone (NVP) copolymer has been the most used sorbent in the isolation of BPA because on the one hand, the hydrophilic NVP copolymer permits good wettability of the sorbent and acts as a hydrogen acceptor, and on the other hand, the hydrophobic divinylbenzene polymer provides reversed-phase retention for BPA. Nevertheless, it is proven that the use of polystyrene–divinylbenzene sorbents does not improve recoveries despite offering hydrophobic and π–π interactions for the retention of BPA [5].
- Selective sorbents: these are highly selective SPE sorbent materials designed for the determination of trace levels of a single analyte or a type of structurally related compound, performing extraction and clean-up in one step. It must be borne in mind that the development of these materials is time-consuming, complex and expensive [5].
- Restricted access materials: these materials permit online clean-up/extraction and act as a barrier that excludes macromolecules like proteins, while low molecular mass analytes are retained by conventional mechanisms, either by physical diffusion barriers (pore diameter) or by a chemical barrier (hydrophobic, ionic or affinity interactions) [5, 29].
- Immunoaffinity columns (also called immunosorbents): these columns are composed of antibodies that can be physically entrapped in the pores of a silica network or covalently bound to a solid support material, which is the most commonly used method. However, immunosorbents produced by immobilisation of the antibodies in the pores of a sol–gel generated silica matrix have several advantages including they are simpler, cheaper, more versatile, reusable and they have higher storage stability [5, 30]. The main drawbacks of this technique are the high cost to produce the antibodies, the short life of columns and food samples have to be treated to make them compatible since organic solvents can cause irreversible denaturation of antibodies [5].
- Molecularly imprinted polymers (MIPs): these are synthetic polymers with molecular recognition ability for a specific analyte or a group of structurally related species. MIPs are synthesised from functional monomers (e.g., methacrylic acid, acrylamide and 4-vinylpyridine) assembled around a template molecule, with a chemical structure similar to the target molecule, and polymerised in the presence of a cross-linker (e.g., ethylene glycol dimethacrylate and trimethylolpropane trimethacrylate) and a suitable solvent [5, 31]. The elimination of the

template molecule then gives a cavity with size, shape and chemical functionality complementary to the target molecule [32]. MIPs offer some advantages such as high chemical stability and resistance against organic solvents, strong acids and bases and heating; they are low cost and easy to synthesise, they permit the use of larger sample volumes and are reusable [5].

Other less used extraction techniques, such as microwave-assisted extraction (MAE), pressurised liquid extraction (PLE), solid-phase microextraction (SPME) and stir-bar sorptive extraction (SBSE), are capable of improving the isolation and clean-up steps in terms of sample size, sample handling reduction, automation, solvent consumption and analysis time. Matrix solid-phase dispersion (MSPD) has the potential to simplify the extraction of solid samples [5].

MAE is a technique that is based on the application of microwave energy to the sample during extraction, which allows it to be agitated and heated quickly. It is a good alternative to solvent extraction from solid and semisolid food samples because it has the advantage of achieving good extraction efficiencies, saving solvent and time [5].

PLE, also known as accelerated solvent extraction, combines elevated temperatures (40–200 °C) and pressures (1,000–2,500 psi) with liquid solvents. Under these conditions, solvents have enhanced solvation power and this enables fast and efficient extraction rates. Optimisation of the extraction process for PLE generally includes choosing the most appropriated solvent type, temperature and pressure. The solvents used for BPA extraction have been DCM for meat, methanol for cereals and acetone-*n*-hexane (1:1, v/v) for fish liver [5].

SPME is based on the distribution of a target analyte which is governed by its affinity between a fused silica fibre coated with an appropriate stationary phase and the sample [33]. The extraction can be performed by suspending the fibre in the vapour phase above the liquid sample [headspace (HS)–SPME], which allows the extraction of volatile or semivolatile compounds from foods; or by direct immersion (DI)–SPME, which needs a thorough clean-up because the reproducibility and robustness of the fibre can be affected by adsorption of other matrix components or clogged by particles. The desorption step can then be carried out thermally by exposing the fibre in the hot injection port of a GC, or chemically in an organic solvent coupled with LC. Different stationary phases were assessed for the extraction of BPA by DI–SPME with the best results achieved using the more polar fibres [5].

SBSE was introduced in 1999 as a sample preparation technique that minimised the use of organic solvents. It could be described as a microextraction method, where the analyte is preconcentrated on a stir bar coated with a polydimethylsiloxane (PDMS) phase. Nonpolar PDMS is the most appropriate available polymer; hence, the application of SBSE to polar and semipolar compounds requires an additional derivatisation step before extraction, such as acetylation and silylation [34]. The stir bar can be immersed into the liquid sample or held in the HS for the extraction of volatile or semivolatile compounds. The removal of analytes from the bar is then achieved by GC thermal desorption or by elution with a solvent when coupled with LC. Another stationary phase for SBSE based on PDMS/β-cyclodextrin allows the extraction of BPA as the cyclodextrin presents a high number of functional hydroxyl groups which enable an increased recovery of polar compounds [5, 35].

MSPD is a sample preparation procedure suited to the extraction of solid, semi-solid and/or highly viscous food that allows simultaneous extraction and clean-up, which results in reduced solvent use and time. In this technique, the sample is mixed with a sorbent (C_{18}-bonded silica, sodium sulphate or diatomaceous earth), followed by washing with a solvent or packaged into an SPE column before elution [5, 36].

Quick, Easy, Cheap, Effective, Rugged and Safe (QuEChERS) is a recently developed extraction technique for food analysis based on a first step of solvent extraction, with acetonitrile and partitioning with magnesium sulphate (alone or in combination with other salts), followed by a second clean-up step using dispersive SPE with the aim of eliminating possible interfering compounds [21, 37]. Its advantages include simplicity, effectiveness, low solvent consumption (resulting in low costs), extraction speed, flexibility and selectivity [21].

2.9 Analytical methods

The determination of BPA in food products requires the use of highly sensitive and selective techniques due to the low levels at which it is frequently found and the complexity of food matrices. Although the SML set by the EU Commission is relatively high (0.6 mg/kg), the reported low dose effects of BPA has necessitated the development of analytical methods with a limit of detection low enough to assess the risk of human exposure at these levels [5]. Presently, BPA is mostly analysed using chromatographic methods such as LC–fluorescence detection (FLD), LC–MS and GC–MS. LC offers the advantage of simplicity since GC requires a time-consuming derivatisation step due to the polar nature and inadequate volatility of BPA for direct analysis, hence introducing new sources of error [3, 5, 33].

LC of BPA is generally carried out using reversed-phase C_{18} columns. The composition of the mobile phase varies depending on the type of detection, for example, water and acetonitrile are most commonly used for FLD, while water and methanol are preferred for MS methods [5, 7]. LC is usually performed at room temperature but high temperatures (up to 40 °C) are sometimes employed because they allow reduced analysis time and increased reproducibility. The run time usually ranges between 15 and 40 min [5]. Some LC methods developed for the analysis of BPA are described in Table 2.2.

Table 2.2: LC methods.

Analytes	Matrix	Detector	Sample preparation or extraction	Column/mobile phase	LOD/LOQ	Results	Reference
BPA	Food-contact recycled paper materials	MS	Focused ultrasonic solid–liquid extraction	Column: BEH C18 (50 mm × 2.1 mm × 1.7 µm) with a precolumn of the same material Mobile phase: 1:1 acetonitrile:methanol mixture (A) and a 0.5 mM sodium acetate 8.5 mM acetic acid aqueous solution (B) Gradient: increased from 5% to 50% A within 1 min, was held at 50% A for 1 min, increased from 50% to 95% A within 2 min, maintained at 95% A for 1 min and then returned to the initial conditions	34 ng/mL (50 ng/mL)	BPA was found in all samples with the highest concentration in paper tablecloth, pizza box and packaging box	[10]
BPA	Breast milk	MS	Solid-phase extraction and derivatisation	Column: BEH C18 (100 mm × 21 mm × 1.7 µm) Mobile phase and gradient: 30% acetonitrile and 70% 0.1% acetic acid held for 0.5 min, increased to 61% acetonitrile from 0.5 to 6.5 min, held under these conditions for 2.5 min and finally reversed to 30% acetonitrile over 1 min, with a 3 min hold between injections	0.22 ng/mL (–)	BPA was detected in 62% of the milk samples with a median concentration of 0.68 ng/mL	[16]

(continued)

Table 2.2 (continued)

Analytes	Matrix	Detector	Sample preparation or extraction	Column/mobile phase	LOD/LOQ	Results	Reference
BPA	20 breast milk samples	MS/MS	Restricted access materials with LiChrospher RP-18 ADS	Column: Chromolith Performance RP-18 (100 mm × 4.6 mm) Mobile phase: water (A) and methanol (B) Gradient: 50% B for 5 min, increased to 65%B from 5 to 13 min, increased to 100% B from 13 to 20 min, kept at 100% B from 20 to 22 min, then decreased to 50% B at 22.1 min, and kept at 50% B for 2 min	0.28 ng/mL (–)	BPAs were frequently detected in the milk samples with a median concentration of free BPA of 0.4 ng/mL and of total BPA of 1.1 ng/mL	[29]
BPA	40 canned energy drinks	FLD	Molecularly imprinted polymer	Column: C18 (75 mm × 4.6 mm × 2.7 µm) Mobile phase: water (A) and acetonitrile (B) Gradient: 50% B for 0.5 min, a linear gradient 50%–90% B in 5.5 min, then holding at 95% B for 3 min, 95%–50% B in 2 min and equilibrated at 50% B for further 4 min	0.15 ng/mL (0.5 ng/mL)	BPA was detected in 17 out of 40 samples (42.5%)	[31]

BPA	72 baby bottle samples (12 different brands)	FLD	SPE	Column: Hypersil C19 (250 mm × 4.6 mm × 5 µm) Mobile phase: water (A) and acetonitrile (B) Gradient: 0–2 min, 25% B; 2–15 min, 25–45% B; 15–38 min, 45–70% B; 38–45 min, 70–100% B; 45–50 min, 100% B; 50–52 min, 25% B	0.004 mg/kg for 3% acetic acid, 0.005 mg/kg for 50% ethanol and 0.007 for infant formula (0.03 mg/kg)	No detection of BPA into the acidic simulant. In four brands, there was detectable migration into the simulant 50% ethanol (v/v) and BPA was detected in only two samples of infant formula	[38]
BPA	10 samples of different powdered milks (2) and infant formulas (8)	MS	Pressurised liquid extraction	Column: Gemini C18 (250 mm × 4.6 mm × 5 µm) Isocratic conditions Mobile phase: 5% water and 95% methanol/ acetonitrile (90/10, v/v), both with 0.1% ammonia	0.005 mg/kg (0.016 mg/ kg)	BPA was present in four infant formulas and in one powdered milk in the range 0.07–1.29 mg/ kg	[39]
BPA	11 canned soft drinks including soda, beer, cola, tea and energy drinks	MS	Online SPE	Column: C18 Fused-Core (50 mm × 2.1 mm × 2.7 µm) Mobile phase: methanol–water Elution program: 0 min 15% methanol; from 0 to 3 min a linear gradient elution up to 80% methanol followed by an isocratic step of 3.5 min	25 ng/L (85 ng/L) for tonic and cola, while 15 ng/L (50 ng/L) for lemon soda	BPA was detected in most of the samples at concentrations ranging from 44 to 607 ng/L	[40]

(continued)

Table 2.2 (continued)

Analytes	Matrix	Detector	Sample preparation or extraction	Column/mobile phase	LOD/LOQ	Results	Reference
BPA	8 bottled water and bottles (6 low-cost bottles and 6 bland baby-feeding bottles)	MS	Without sample clean-up or enrichment	Column: RSLC (50 mm × 2.1 mm × 2.2 µm) Mobile phase: methanol (A) and deionised water (B) Gradient program: A/B (75:25) for 0.1 min, linear gradient to A/B (95:5) in 0.8 min and hold for 2.1 min, then return to initial condition and equilibrate for 2 min	40 ng/L (–)	BPA was not detectable in any of the bottled water samples. Low levels of BPA were found in three bottles after heating in a microwave oven	[41]
BPA	13 kinds of beverages including mineral water, pure drinking water and soda beverages	MS/MS	SPE	Column: Symmetry C18 (150 mm × 2.1 mm × 3.5 µm) Mobile phase: methanol and water with 0.1% ammonia Gradient: methanol was linearly increased from 10% to 55% in 10 min, then increased to 85% in 10 min and held for 7.5 min, finally brought back to 10% and held for 15 min	0.01 ng/L (0.04 ng/L) for mineral drinking water and 0.6 ng/L (2.0 ng/L) for soda beverages	BPA was not detected in any sample	[42]

Note: (–), not available.
LOD, limit of detection; LOQ, limit of quantification; MS, mass spectrometry; FLD, fluorescence detection; SPE, solid-phase extraction.

BPA exhibits native fluorescence with an excitation wavelength at 275 nm and an emission wavelength at 305 nm. The identification of BPA in the sample is based on the retention times, so the possibility of interferences from other fluorescent migrants present in the sample that appear simultaneously at the same retention time should always be considered because they may produce false-positives. To overcome this, later confirmation using LC–MS or GC–MS has sometimes been employed [5].

MS allows the determination of a certain analyte by the selection of specific ions or transitions. In comparison with FLD, MS offers higher confidence in peak identification as it is based on the mass spectrum. Suitable sample preparation is necessary as the presence of matrix components can affect the ionisation efficiency and produce background noise. The analysis of BPA by LC–MS can be carried out using different types of ionisation sources: atmospheric pressure ionisation interfaces, namely electrospray ionisation (ESI) or atmospheric pressure chemical ionisation, both in negative mode. Between the two, ESI is more frequently used because it generally provides better sensitivity. The response of ESI–MS for the determination of' BPA is strongly influenced by the mobile-phase composition; it has been found that methanol–water gives a higher response for BPA than acetonitrile–water owing to the lower boiling point of the former, which favours the desolvation of the electrospray droplets. The response using acetonitrile–water can be increased upon the addition of modifiers, such as 0.5% and 0.01% of ammonia or 0.01% of acetic acid [5].

Independently of the type of ionisation source, the most abundant ion in the BPA mass spectrum, which is used for quantitation purposes, is $[M–H]^-$ *m/z* 227. In LC–tandem MS (MS/MS), for confirmation and quantification of BPA, one or more MS/MS transition reactions of the precursor ion *m/z* 227 are monitored. For example, transition *m/z* 227 > 211 is related to the additional loss of oxygen $[M–H–O]^-$, transition *m/z* 227 > 212 (the most prominent product ion obtained) is related to the additional loss of a methyl radical $[M–H–CH_3]^{\bullet -}$, transition *m/z* 227 > 113 (using acetonitrile – 0.01% ammonia in water as the mobile phase) is related to the additional loss of both acidic protons $[M–2H]^{2-}$, transition *m/z* 227 > 133 results from the addition cleavage of a phenol group $[M–H–C_6H_5OH]^-$ *m/z* 133 and transition *m/z* 227 > 93 is formed by the additional loss of hydroxyphenyl propyl $[M–H–C_9H_{10}O]^-$ [5].

Most MS methods include the addition of an internal standard to compensate for the potential loss of analyte during sample preparation, matrix effects and for any variation in instrument performance [3, 5]. In the case of the most common internal standard, BPA–d_{16} (molecular mass 244), the most abundant ion in the mass spectrum, is *m/z* 241 $\{[(M–D_2+H_2)–H]^-\}$, owing to the transformation into BPA–d_{14} (molecular mass 242), as the two acidic deuterium atoms are exchanged for protons when dissolved in a protic medium. The product ions that are monitored are *m/z* 223, which is attributed to the loss of the CD_3 radical and *m/z* 142 formed by the loss of a phenol–d_4 group together with a deuterium atom derived from a CD_3 group [5]. While the most abundant ion for the isotopically labelled internal standard $^{13}C_{12}$–BPA is ([M-H]$^-$), the signal assigned to *m*/*z* 239 [43].

GC–MS (with electron impact) provides a higher resolution and sensitivity (lower detection limits) than LC–MS for the determination of BPA, although it usually requires a derivatisation step, which introduces new sources of error, mainly due to contamination. The electron mass spectra of BPA is characterised by a molecular ion at *m*/*z* 228 ($[C_{15}H_{13}O_2]^+$) with the most abundant fragment ion at *m*/*z* 213, which corresponds to the loss of a methyl group ($[C_{14}H_{13}O_2]^+$). Another minor fragmentation route involves the loss of one of the aryl groups to give a tert-benzylic carbocation ($[C_9H_8O]^+$) at *m*/*z* 135 and the subsequent loss of methane to give a fragment ion at *m*/*z* 119 [5]. Quantitation requires a previous derivatisation step for the phenolic hydroxyl groups of BPA, and several derivatisation procedures have been reported, including silylation, acetylation and alkylation reactions [33]. Silylation is the main derivatisation method used, with the most commonly used agent N–O–bis(trimethylsilyl)trifluoroacetamide containing 1% of trimethylchlorosilane, which favours the formation of the single derivative. The electron mass spectra of the bis(trimethylsilylether) derivative of BPA is characterised by a molecular ion at *m*/*z* 372 ($[C_{21}H_{32}Si_2O_2]^+$) with the most abundant fragment ion at *m*/*z* 357, which corresponds to the direct loss of a methyl group ($[C_{20}H_{29}Si_2O_2]^+$) [5]. However, silylation requires temperature control and nonaqueous media, and is time-consuming [44]. For acetylation of the hydroxyl groups of BPA, the derivatisation agents commonly used are acetic anhydride and trifluoroacetic anhydride. The electron mass spectra of the *O*-bis(trifluoroacetyl) derivative of BPA is characterised by a molecular ion at *m*/*z* 420 and the most abundant fragment ion at *m*/*z* 405, which corresponds to the direct loss of a methyl group. The *O*-bis(trifluoroacetyl) derivative was more sensitive than the trimethylsilyl derivative due to the higher molecular mass, which led to more favourable signal-to-noise ratios, and the monoisotopic nature of fluorine that prevented a reduction of ion intensity. In GC–MS, as in LC–MS, the use of internal standards, such as deuterated BPA–d_{16} and BPA–d_{14}, is common [5]. Some GC methods developed for the analysis of BPA are described in Table 2.3.

Other detection techniques, such as ultraviolet, electrochemical or immunochemical have been reported but to a lesser extent. Immunochemical methods, like the enzyme-linked immunosorbent assay, have several advantages as it is easy and rapid to perform, provides good sensitivity and specificity and doesn't require qualified personnel or expensive equipment. BPA is not able to initiate an immune response itself due to its small size, so it needs to be conjugated with a protein [5].

Table 2.3: GC methods.

Analytes	Matrix	Detector	Sample preparation or extraction	Column/injection mode/ oven temperature	LOD/(LOQ)	Results	Reference
BPA	105 samples (canned and plastic wrapped food for human consumption; fresh meat and fish; and canned and plastic wrapped pet food)	Mass spectrometry (MS)	Extraction of the sample with acetonitrile in an ultrasonic bath at 40 °C. Liquid/liquid extraction with hexane and purification on ENVI-Carb using hexane. Derivatisation with BSTFA and purification on silica column	Column: DB-5MS (30 m × 0.25 mm × 0.25 μm) Injection mode: Splitless Oven temperature: 120 °C for 1 min, 10 °C/min to 240 °C, and 30 °C/min and hold 10 min	0.20 ng/g wet weight (–)	BPA was found in 63 samples. BPA concentrations were higher for foods with pH 5 than more acidic and alkaline foods	[12]
BPA	47 canned seafood samples	MS	Dispersive solid-phase clean-up (QuEChERS) with dispersive liquid–liquid microextraction (DLLME) and derivatisation with acetylating agents	Column: DB-5MS (30 m× 0.25 mm × 0.25 μm) Injection mode: Splitless Oven temperature: 100 °C held for 1 min, ramped to 280 °C at 30 °C/min held for 5 min	0.2 μg/kg (1 μg/kg)	The presence of BPA was confirmed in more than 83% of the samples (1.0–99.9 μg/kg)	[21]

(continued)

Table 2.3 (continued)

Analytes	Matrix	Detector	Sample preparation or extraction	Column/injection mode/ oven temperature	LOD/(LOQ)	Results	Reference
BPA	Six polycarbonate bottles	MS	Ethyl chloroformate derivatisation for solid-phase microextraction (SPME)	Column: VF-5MS capillary column (30 m × 0.25 mm × 0.25 µm) Injection mode: Splitless Oven temperature: 100 °C (1 min); 20 °C/min to 210 °C (1 min); 45° C/min to 300 °C (10 min)	0.1 µg/L (0.38 µg/L) for milk and 0.01 µg/L (0.052 µg/L) for water	BPA level migrated into water samples was found to be in the range of not detected–70 µg/L	[33]
BPA	Vegetables (carrots, lettuce, onions and tomatoes)	MS/MS	Matrix solid-phase dispersion	Column: ZB-5MS (30 m × 0.25 mm × 0.25 µm) Injection mode: Splitless Oven temperature: 50 °C (held for 0.5 min), increased at 20 °C/min to 300 °C where it was held 2.5 min (total analysis time was 15.5 min)	0.1 ng/g for onion (0.4 ng/g) and tomato (0.2 ng/g), and 0.2 ng/g for lettuce (0.8 ng/g) and carrot (0.7 ng/g)	BPA was detected in all the samples at concentrations up to 16 ng/g	[43]

BPA	78 canned food products (pasta, vegetable, tomato paste, soup and tuna products)	MS	Solid-phase extraction followed by derivatisation	Column: HP-5MS (30 m × 0.25 mm × 0.25 µm) Injection mode: Splitless Oven temperature: 100 °C for 1 min, raised to 225 °C for 5 min at 20 °C/min, raised to 325 °C at 35 °C/min and held for 1 min	0.60 ng/g (–)	BPA was detected in almost all 78 canned food products; the BPA concentration in only one product (tomato paste) was below the LOD. Canned tuna products had the highest BPA level (534 ng/g)	[45]
BPA	12 polycarbonate baby bottles	MS	SPE	Column: Zebron ZB-5 (30 m × 0.25 mm × 0.25 µm) Injection mode: Splitless Oven temperature: 4 min at 40 °C, then at 10 °C/min to 300 °C and held for 10 min	0.1 µg/L (1 µg/L)	The mean BPA level from new bottles was 0.23 ± 0.12 µg/L, while from bottles subjected to simulated use were 8.4 ± 4 µg/L (dishwashed 51 times) and 6.7 ± 4 µg/L (dishwashed 169 times)	[46]
BPA	Tap water, bottled water and baby bottles	MS	SPE	Column: DB5-MS (30 m × 0.25 mm × 0.25 µm) Injection mode: Splitless Oven temperature: from 80 °C (1 min) to 220 °C at 10 °C/min, from 220 °C to 260°C at 4 °C/min, and from 260 °C to 300 °C (8 min) at 5 °C/min, then to 310 °C (15 min) at 20 °C/min	0.7 ng/L (2.0 ng/L)	17 out of 21 bottled water contained BPA (17.6–324 ng/L) and all of the tap waters contained BPA (2.3–317 ng/L). BPA was found in 4 brands of baby bottles at room temperature of 24 °C (267 ng/L), 40 °C (987 ng/L) and 100 °C (4,500 ng/L)	[47]

(continued)

Table 2.3 (continued)

Analytes	Matrix	Detector	Sample preparation or extraction	Column/injection mode/ oven temperature	LOD/(LOQ)	Results	Reference
BPA	Vegetables and fruits	MS	Acidic hydrolysis to release the conjugated form and trimethylsilyl derivatisation	Column: Rxi-5MS (30 m × 0.25 mm × 0.25 μm) Injection mode: Splitless Oven temperature: initial temperature, 130 °C; increased to 280 °C at 10 °C/min; held at 280 °C for 5 min	0.03–0.3 μg/kg (0.1–1.0 μg/kg)	BPA was detected in all vegetable and fruits, ranging from 0.2 ± 0.1 to 9.0 ± 4.9 μg/kg	[48]
BPA	Canned vegetables and fruits	MS	Dispersive solid-phase clean-up (QuEChERS) with dispersive liquid–liquid microextraction (DLLME)	Column: DB-5MS (30 m × 0.25 mm × 0.25 μm) Injection mode: Splitless Oven temperature: 100 °C held for 1 min, ramped to 280 °C at 30 °C/min held for 5 min	0.3 μg/kg (1 μg/kg)	Presence of BPA in more than 87% (34 out of 39) of the samples (3.7–265.6 μg/kg)	[49]

References

[1] A. Careghini, A.F. Mastorgio, S. Saponaro and E. Sezenna, *Environmental Science and Pollution Research*, 2015, 22, 8, 5711.

[2] J. Michałowicz, *Environmental Toxicology and Pharmacology*, 2014, 37, 2, 738.

[3] F.M. Ghazali and W.L.W. Johari, *Journal of Biochemistry, Microbiology and Biotechnology*, 2015, 3, 2, 30.

[4] J. Maia, J.M. Cruz, R. Sendón, J. Bustos, M.E. Cirugeda, J.J. Sanchez and P. Paseiro, *Food Research International*, 2010, 43, 5, 1283.

[5] A. Ballesteros-Gómez, S. Rubio and D. Pérez-Bendito, *Journal of Chromatography A*, 2009, 1216, 3, 449.

[6] S.H. Nam, Y.M. Seo and M.G. Kim, *Chemosphere*, 2010, 79, 9, 949.

[7] X.L. Cao and J. Corriveau, *Journal of Agricultural and Food Chemistry*, 2008, 56, 15, 6378.

[8] J. Maia, J.M. Cruz, R. Sendón, J. Bustos, J.J. Sanchez and P. Paseiro, *Food Research International*, 2009, 42, 10, 1410.

[9] E. Maiolini, E. Ferri, A.L. Pitasi, A. Montoya, M. Di Giovanni, E. Errani and S. Girotti, *Analyst*, 2014, 139, 1, 318.

[10] D. Pérez-Palacios, M.A. Fernández-Recio, C. Moreta and M.T. Tena, *Talanta*, 2012, 99, 167.

[11] J. López-Cervantes and P. Paseiro-Losada, *Food Additives & Contaminants*, 2003, 20, 6, 596.

[12] A. Schecter, N. Malik, D. Haffner, S. Smith, T.R Harris, O. Paepke and L. Birnbaum, *Environmental Science & Technology*, 2010, 44, 24, 9425.

[13] T.T. Schug and L.S. Birnbaum in *Toxicants in Food* Packaging *and Household Plastics: Exposure and Health Risks to Consumers*, Ed., S. M. Snedeker, Springer-Verlag, London, UK, 2014, p.1.

[14] C. Kubwabo, I. Kosarac, B. Stewart, B.R. Gauthier, K. Lalonde and P.J. Lalonde, *Food Additives & Contaminants*, 2009, 26, 6, 928.

[15] S.A. Baluka and W.K. Rumbeiha, *Food and Chemical Toxicology*, 2016, 92, 58.

[16] S.M. Zimmers, E.P. Browne, P.W. O'Keefe, D.L. Anderton, L. Kramer, D.A. Reckhow and K.F. Arcaro, *Chemosphere*, 2014, 104, 237.

[17] M.V. Maffini, B.S. Rubin, C. Sonnenschein and A.M. Soto, *Molecular and Cellular Endocrinology*, 2006, 254, 179.

[18] B. Yi, C. Kim and M. Yang, *Journal of Chromatography B*, 2010, 878, 27, 2606.

[19 European Food Safety Authority CEF Panel (Panel on Food Contact Materials, Enzymes, Flavourings and Processing Aids), *EFSA Journal*, 2015, 13, 1, 3978.

[20] W. Völkel, T. Colnot, G.A. Csanády, J.G. Filser and W. Dekant, *Chemical Research in Toxicology*, 2002, 15, 10, 1281.

[21] S.C. Cunha, C. Cunha, A.R. Ferreira and J.O. Fernandes, *Analytical and Bioanalytical Chemistry*, 2012, 404, 8, 2453.

[22] European Commission draft: Ref. Ares (2017) 4140854 – 23/ 08/2017:*Bisphenol A in Varnishes and Coatings and Plastics Intended to Come into Contact with Food*. https://ec.europa.eu/info/law/better-regulation/initiatives/ares-2017-4140854_en

[23] E. Fasano, F. Bono-Blay, T. Cirillo, P. Montuori and S. Lacorte, *Food Control*, 2012, 27, 1, 132.

[24] J. Muncke in *Toxicants in Food Packaging and Household Plastic: Exposure and Health Risks to Consumers*, Ed., S.M. Snedeker, Springer-Verlag, London, UK, 2014, p.265.

[25] M.I. Santillana, E. Ruis, M.T. Nieto, A.R.B. De Quirós, R. Sendón, M.E. Cirugeda and J.J. Sanchez, *European Food Research and Technology*, 2013, 236, 5, 883.

[26] N. Milić, D. Četojević-Simin, M. Milanović, J. Sudji, N. Milošević, N. Ćurić, L. Abenavoli and M. Medić-Stojanoska, *Food and Chemical Toxicology*, 2015, 83, 268.

[27] R.A. Rudel, J.M. Gray, C.L. Engel, T.W. Rawsthorne, R.E. Dodson, J.M. Ackerman, J. Rizzo, J.L. Nudelman and J.G. Brody, *Environmental Health Perspectives*, 2011, 119, 7, 914.

[28] H. Gallart-Ayala, O. Núñez and P. Lucci, *Trends in Analytical Chemistry*, 2013, 42, 99.

[29] X. Ye, Z. Kuklenyik, L.L. Needham and A.M. Calafat, *Journal of Chromatography B*, 2006, 831, 1, 110.

[30] M. Cichna-Markl, *Methods*, 2012, 56, 2, 186.

[31] P. Gallo, I.D.M. Pisciottano, F. Esposito, E. Fasano, G. Scognamiglio, G.D. Mita and T. Cirillo, *Food Chemistry*, 2017, 220, 406.

[32] Y. Wen, L. Chen, J. Li, D. Liu and L. Chen, *Trends in Analytical Chemistry*, 2014, 59, 26.

[33] M.K.R. Mudiam, R. Jain, V.K. Dua, A.K. Singh, V.P. Sharma and R.C. Murthy, *Analytical and Bioanalytical Chemistry*, 2011, 401, 5, 1695.

[34] A. Iparraguirre, A. Prieto, P. Navarro, M. Olivares, L.A. Fernández and O. Zuloaga, *Analytical and Bioanalytical Chemistry*, 2011, 401, 1, 339.

[35] Y. Hu, Y. Zheng, F. Zhu and G. Li, *Journal of Chromatography A*, 2007, 1148, 1, 16.

[36] B. Albero, C. Sánchez-Brunete, E. Miguel and J.L. Tadeo, *Food Chemistry*, 2017, 217, 660.

[37] M.Á. González-Curbelo, B. Socas-Rodríguez, A.V. Herrera-Herrera, J. González-Sálamo, J. Hernández-Borges and M.Á. Rodríguez-Delgado, *Trends in Analytical Chemistry*, 2015, 71, 169.

[38] M.I. Santillana, E. Ruiz, M.T. Nieto, J. Bustos, J. Maia, R. Sendón and J.J. Sánchez, *Food Additives & Contaminants: Part A*, 2011, 28, 11, 1610.

[39] E. Ferrer, E. Santoni, S. Vittori, G. Font, J. Mañes and G. Sagratini, *Food Chemistry*, 2011, 126, 1, 360.

[40] H. Gallart-Ayala, E. Moyano and M.T. Galceran, *Analytica Chimica Acta*, 2011, 683, 2, 227.

[41] J. Wang and W.C. Schnute, *Rapid Communications in Mass Spectrometry*, 2010, 24, 17, 2605.

[42] B. Shao, H. Han, J. Hu, J. Zhao, G. Wu, Y. Xue, Y. Ma and S. Zhang, *Analytica Chimica Acta*, 2005, 530, 245.

[43] K. Inoue, M. Kawaguchi, Y. Funakoshi and H. Nakazawa, *Journal of Chromatography B*, 2003, 798, 1, 17.

[44] Q. Wang, L. Zhu, M. Chen, X. Ma, X. Wang and J. Xia, *Chemosphere*, 2017, 169, 709.

[45] X.L. Cao, J. Corriveau and S. Popovic, *Journal of Food Protection*, 2010, 73, 6, 1085.

[46] C. Brede, P. Fjeldal, I. Skjevrak and H. Herikstad, *Food Additives & Contaminants*, 2003, 20, 7, 684.

[47] X. Li, G.G. Ying, H.C. Su, X.B. Yang and L. Wang, *Environment International*, 2010, 36, 6, 557.

[48] J. Lu, J. Wu, P.J. Stoffella and P.C. Wilson, *Journal of Agricultural and Food Chemistry*, 2012, 61, 1, 84.

[49] S.C. Cunha and J.O. Fernandes, *Food Control*, 2013, 33, 2, 549.

3 Melamine

3.1 Introduction

Melamine (2,4,6-triamino-1,3,5-triazine) is a heterocyclic triazine that contains three amino groups [1–3]. This small polar molecular compound is found in the form of colourless monoclinic prisms, and its solubility in water and organic solvents is sparse [4].

Melamine was first developed in the 1830s by Justus von Liebig, a German chemist, by heating a mixture of potassium thiocyanate and ammonium chloride [4, 5]. It was subsequently obtained using numerous methods involving heating different starting materials including guanidine, carbonate, thiourea, cyanamide or dicyandiamide [4]. Commercially produced melamine is manufactured using urea as a starting material. There are some inconsistencies in the literature regarding the manufacture of melamine from dicyandiamide. Some sources indicate that commercial production of melamine from the thermal condensation of dicyandiamide ceased during the 1980s, while other sources indicate that this process is still used to manufacture melamine [6].

Melamine can be degraded via deamination reactions to form the by-products ammeline (4,6-diamino-2-hydroxy-1,3,5-triazine), ammelide (6-amino-2,4-dihydroxy-1,3,5-triazine) and cyanuric acid (2,4,6-trihydroxy-1,3,5-triazine) by industrial synthesis, under strong acidic and alkaline conditions, or through microbial degradation using two strains of soil bacteria, *Pseudomonas* strain A and *Klebsiella terragena*. The melamine metabolic breakdown pathway leads to the formation of ammeline, then ammelide and finally, cyanuric acid, which is further broken down into biuret, urea and ultimately, ammonia and carbon dioxide (CO_2) [3, 6, 7].

Cyanuric acid is an oxytriazine, while ammelide and ammeline are, respectively, monoamino- and diaminooxytriazine analogues. Ammeline can be synthesised either by heating dicyandiamide with aqueous ammonia at 160–170 °C or by heating melamine with concentrated sulfuric acid for a brief period at 190 °C. Ammelide decomposes at 170 °C with water to form CO_2 and ammonia [3]. The physical and chemical properties of melamine and its analogues are listed in Table 3.1.[6]

3.2 Industrial applications of melamine and its analogues

Melamine is an industrial chemical used primarily in conjunction with formaldehyde for the synthesis of melamine formaldehyde (MF) resins. Such resins are used for the manufacture of many plastics, laminates, coatings, commercial filters, glues and adhesives. Melamine resins are also used in a wide range of flame-resistant

https://doi.org/10.1515/9783110648065-003

Table 3.1: Physical and chemical properties of melamine and its analogues [6].

	Melamine	**Cyanuric acid**	**Ammelide**	**Ammeline**
Structure	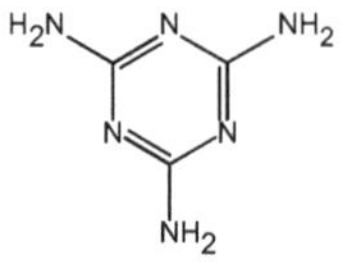	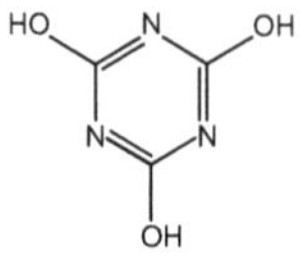	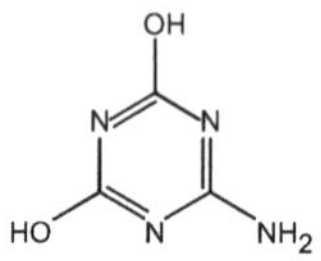	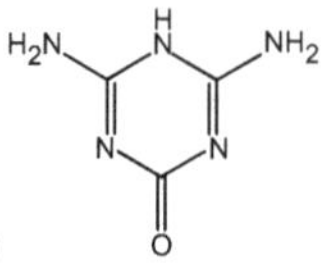
Chemical formula	$C_3H_6N_6$	$C_3H_3N_3O_3$	$C_3H_4N_4O_2$	$C_3H_5N_5O$
CAS number	108-78-1	108-80-5	645-93-2	645-92-1
Molecular weight (g/mol)	126.12	129.07	128.09	127.10
% nitrogen (w/w)	66.6	32.6	43.7	55.1
Appearance	Fine white crystalline powder	White crystalline solid	White powder	White powder
Melting point (°C)	345-347 Descomposes	360	Descomposes	Descomposes
Aqueous solubility (mg/L)	3240 (at 20°C)	2000 (at 25°C)	76.9	75
Pk_a (dissociation constant)	5.35 (at 25°C)	4.74 (at 25°C)		9.65 (at 40°C)

materials as a result of their fire-retardant properties due to releasing nitrogen gas when burned or charred [4, 7].

Melamine is not a natural product and is not approved for direct addition to any human or animal food; in fact nor are there any guidelines in the *Codex Alimentarius* [4, 7]. However, melamine can be present at low levels in food due to its legal use in food contact materials (FCM) [8]. The MF polymer is considered suitable for food contact applications due to certain properties, including hardness, heat resistance and general stability. Melamine is a common monomer in the manufacture of plastics and is predominantly used in the production of tableware such as cups, bowls, plates and utensils [3]. However, melamine can migrate from plastic tableware into food, especially at high temperature, acidic pH and upon repeated use [9]. For this reason, tableware made of melamine should not be exposed to elevated temperatures (around 120 °C) such as in the oven or microwave [3, 4]. According to the European Food Safety Agency (EFSA), articles made of melamine plastic are considered to be the most important source of exposure to melamine from FCM [9].

Melamine can also be formed as a metabolite of cyromazine, a pesticide or veterinary drug used to prevent flies from hatching. Residues of cyromazine and melamine have been detected on vegetable crops [7, 10].

Trichloromelamine, which can decompose to melamine, is permitted for use in the United States in sanitising solutions for food-processing equipment, utensils and food contact articles (except milk containers or equipment) [3, 7].

Ammeline, ammelide and cyanuric acid are not included on the European Union positive list of substances intended to be used in food contact plastics [9]. Ammeline is used in lubricating greases and no information regarding the uses of ammelide is available [6]. Trace levels of cyanuric acid can be present in food and water from the use of dichloroisocyanurate (a disinfectant used in water treatment) in drinking water, swimming pool water and water used in food manufacturing [7].

3.3 Fraudulent uses

Although this section is not directly related to FCM, it is included and discussed due to its importance. As a result of increasing globalisation of the international food market, foodstuffs can be spread around the planet all too quickly and pose a threat to practically everyone, but particularly to vulnerable groups such as infants, young children, pregnant women, the elderly and those already ill [7, 8]. Melamine is a nitrogen-rich industrial chemical that has been fraudulently added to edible foodstuffs in order to give a false appearance of a high level of protein and so reduce costs [1, 2, 11, 12].

Melamine is not distinguishable from protein when measured using the Kjeldahl and Dumas test, thus giving falsely high readings [7]. These nonspecific procedures cannot distinguish between nitrogen incorporated into proteins, which is present in amino acids and therefore in proteins, and nonprotein nitrogen present in many other nonproteinaceous molecules such as melamine [2, 4]. These methods rely on the liberation of ammonia from the protein by concentrated sulfuric acid or heat treatment, respectively [13].

Since 1958, melamine has been added to cattle feed as a nonprotein nitrogen source. However, its use was discontinued in 1978 because it was found to be completely hydrolysed in ruminants. In March 2004, an outbreak of food adulteration involving melamine led to renal failure in dogs and cats in several Asian countries due to consumption of pet food products manufactured at a single factory in Thailand [3, 12].

In March 2007, pet food contaminated with melamine, and melamine-related compounds, resulted in an outbreak of renal disease and even associated deaths in cats and dogs in the United States [1]. Further investigation showed that some wheat gluten and rice protein extracts imported from China as pet food ingredients were contaminated with melamine and its analogues, and these compounds are not approved for use as ingredients in pet food [3, 7]. This adulteration caused hundreds of pet deaths due to melamine–cyanurate crystals obstructing renal tubules within the kidney causing kidney damage [11].

In July 2008, an abnormally high prevalence of renal stones in infants was discovered in China; however, no action was taken because the authorities did not want a scandal prior to the Beijing Olympics, which were held in August [7]. In September 2008, dairy companies in China were selling milk and infant formula that were contaminated with melamine to falsely increase its protein concentration [1]. Owing to China's wide range of export food products, the contaminated products reached countries on all continents and the situation became an international health scare [1, 7]. Melamine combined with endogenous uric acid to form calculi in the kidney, ureter or bladder [11]. As a result of this incident, about 54,000 infants and young children were hospitalised and at least six children died [8, 12]. Melamine was found in a wide variety of milk and milk products at varying concentrations, from low parts per billion (ppb) to parts per million (ppm) [7]. The Chinese Administration of Quality Supervision, Inspection and Quarantine was reported to have found levels of melamine in dairy products reaching up to 6,200 mg/kg [2].

In addition to its catastrophic health effects, the contamination also had economic effects because the United States and other countries banned the importation of milk and other food and feed products from China [5, 7].

3.4 Metabolism of melamine

For ethical reasons, the absorption, metabolism, distribution, excretion and toxicokinetics of melamine have been systemically investigated in various mammalian species such as mice, dairy goats and pigs [3].

It is assumed that melamine and its metabolites are absorbed in the gastrointestinal tract and systematically distributed [4, 7]. The distribution is probably limited to the body water fraction because melamine is unlikely to be bound in significant amounts to body tissues [3].

Melamine cannot be metabolised by the body; thus, approximately 90% of ingested melamine is excreted within 24 h in its original form by the kidneys in the urine [3, 13]. Melamine may, in a dose-dependent manner, form crystals with either endogenous uric acid or a structural analogue of melamine, cyanuric acid, in renal tubules, leading to progressive tubular blockage and degeneration [4, 7, 8]. When the amount of melamine ingested exceeds the excretion capability of the kidneys, renal disease and even death may occur [3].

3.5 Toxicity

Direct contact with melamine results in skin irritation, and its inhalation causes irritation in the respiratory tract [3]. For ethical reasons, there are no studies regarding the oral toxicity of melamine in humans; the available data are from animal feeding studies [7]. Melamine has been reported to have low toxicity, with lethal doses from oral intake in mice and rats exceeding 3,000 mg/kg body weight. Oral ingestion usually affects the digestive tract, resulting in nausea, vomiting and diarrhoea; and in severe cases, hypertension, oedema or oliguria [3].

Information regarding the toxicity of the analogues, compared with melamine, is limited, and so it is prudent to assume that these analogues have equal effects [7]. Although melamine and cyanuric acid alone are considered to have low human toxicity, the exposure to both may confer a higher risk [5, 7]. It has been demonstrated that the simultaneous exposure to both melamine and cyanuric acid is more toxic as they are able to form an insoluble hydrogen-bonded bimolecular network via a surface-based self-assembly process [1, 3, 7]. In this process, the hydrogen bonding causes crystallisation, which can lead to tissue injury including urolithiasis, bladder cancer or even death [1, 3]. When both melamine and cyanuric acid were administered orally, the most commonly observed effects included body weight loss, bladder and kidney stones, crystalluria and epithelial hyperplasia of the bladder [4]. On the other hand, neither ammeline nor ammelide alone produced any renal effects, but the mixture can produce significant renal damage and crystals in the nephrons [7].

It is known that the kidney is the major target organ of melamine toxicity due to its rapid renal clearance. Therefore, it is clinically important to have a simple method that can reliably detect urinary melamine with an acceptable sensitivity, specificity, precision and accuracy in the routine chemistry laboratory [8].

Melamine, in addition to its nephrotoxicity, is known for its neurotoxicity and causes cognitive impairment and hippocampal injury via adenosine triphosphate depletion; in addition, it also inhibits certain haem groups containing enzymes and activate others [13].

Melamine is associated with carcinogenicity, especially of the renal tract [4, 8]. After oral administration, it has been reported that melamine led to urinary bladder and ureteral carcinomas in male rats and urinary bladder hyperplasia in male mice. High and continuous dietary exposure to melamine is associated with an increase in the production of bladder stones, which is correlated directly with an increased incidence of urinary bladder tumours [4]. Melamine is classified as a group 2B substance according to the World Health Organization (WHO) International Agency for Research on Cancer, classified as ‘possibly carcinogenic to humans’. There is sufficient evidence in experimental animals for melamine to be designated as a carcinogen; however, there is inadequate evidence for carcinogenicity in humans [7].

Melamine is harmful to infants who, as a consequence of the exposure, suffer retarded growth, urinary stone formation and renal failure. Although the mortality rates of infants with melamine-related disease are relatively low, the incidence of stone formation is quite high. Being below average height and underweight are common among children with melamine-related urinary stones because melamine may interfere with the metabolism of glucose, protein and nitrogen in the liver [8].

The possible consequences for pregnant women who are exposed to melamine-contaminated foods is a prime concern. The effects of repeated administration of high doses of triazines, such as melamine and cyanuric acid, in pregnant rats were investigated to determine the risk of these compounds for both mother and foetus. Melamine was found in the serum and amniotic fluid, confirming the potential for transplacental transfer of melamine from mammalian mother to foetus. The presence of melamine in serum is most likely due to the fact that crystals obstruct the renal tubules and effectively impeded excretion. High levels of melamine induce maternal toxicity and also have a negative impact on foetal development, for example, decreased size, and below average foetal body weight and average crown rump length. Studies suggest that the developmental reproductive effects were likely due to renal failure in the mothers, not by the direct effects of melamine on the developing foetus [11]. Lactational transfer of melamine from mother to neonates has been reported in animals [13].

Yet, the long-term health effects of melamine remain unknown [7].

3.6 Regulation

After the reported outbreaks of melamine-contaminated food, the level of concern for food safety increased. Authorities and other organisations established a tolerable daily intake (TDI) for melamine in order to ensure a safe food supply and protect the health of the general population, including infants. This TDI is an estimation of the maximum exposure to a certain agent that the population may be exposed to without any large risk. The US Food and Drug Administration was the first organisation to provide the initial TDI for melamine of 0.63 mg/kg based on animal toxicity assays [14]. The WHO and the EFSA have established a new TDI of 0.2 mg/kg/day for melamine [8]. This TDI is considered appropriate for infants, except for premature babies because they have higher urinary uric acid levels and greater immaturity of kidney function [14].

The TDI for melamine is not applicable if there is a simultaneous exposure with any of its analogues, that is, cyanuric acid, ammelide or ammeline, due to the increased potential for the formation of urinary crystals. A TDI for cyanuric acid of 1.5 mg/kg was proposed based on a previous evaluation of the disinfectant dichloroisocyanurate. However, this TDI makes reference to sodium cyanurate, and its equivalent for cyanuric acid is 1.3 mg/kg, which is a more accurate value, but is not toxicologically different from 1.5 mg/kg. The toxicological databases for ammelide and ammeline are extremely limited and therefore they do not have an established TDI [14].

Melamine is not approved for direct addition to any human or animal food, but melamine can migrate to food due to its legal use as a monomer and additive in plastics used in FCM [8, 14]. For this reason, it is important to monitor the level of melamine in human foods. As a result, governments around the world have set maximum acceptable limits for melamine in food products [15]. In China and the United States, the maximum residue levels for infant formula are 1.0 and 2.5 mg/kg for milk and other milk products (calculated on the basis of ingestion by a person weighing 60 kg), while in the European Regulation on plastics intended to come into contact with foodstuffs, melamine is authorised as a monomer in plastics with a specific migration limit of 2.5 mg/kg [1, 3, 9]. Those products containing more than 2.5 ppm should be destroyed [4]. These limits for melamine provide a sufficient margin of safety for dietary exposure relative to the TDI [7].

3.7 Analysis

Melamine detection and quantification in food and biologic samples has become of considerable interest after the contamination incidents [8]. In order to achieve this, there is a need to establish highly selective and rapid methods that are capable of screening samples and confirming the presence and quantities of melamine and other metabolites [1]. The ideal technique should involve minimal sample preparation, allow sampling under atmospheric conditions and provide a response within seconds of sample introduction [15].

Methods used for the analysis of food for human consumption have been developed or adapted from some methods that were initially used for the study of the pesticide cyromazine, of which melamine is a degradation product. Nowadays, there are various methods available for the analysis of melamine and its structurally related compounds, including cyanuric acid, ammeline and ammelide, in many foods and animal feeds. These techniques can be focused on melamine alone or simultaneously analyse melamine and cyanuric acid in one process. Special care must be taken not to overextend the applicability of a method to a matrix that has not been validated without previous verification and optimisation. Multiresidue methods were also created to analyse melamine residues in animal tissues for toxicology and pathology studies. Methods were also developed for the evaluation of the migration of melamine from MF resin frequently used to make FCM [2].

Currently, liquid chromatography (LC) and gas chromatography (GC) conjugated with mass detection are the most powerful confirmation methods to analyse the presence of melamine [1]. The good selectivity of these methods is due to the use of chromatography to separate analytes from coextracted materials, and the use of multiple reaction monitoring during detection [2]. However, these methods can be expensive and time consuming, and often require sample pretreatment to extract analytes and eliminate matrix effects [1].

With regard to the detection methods used, tandem mass spectrometry (MS/MS) provides the highest degree of selectivity, followed by single-stage mass spectrometry (MS), diode array detection (DAD) and, lastly, ultraviolet (UV) absorption

[1, 2]. The UV and DAD-based methods are less selective and generally less sensitive than the MS methods because they monitor wavelengths in the 200–270 nm region where many organic compounds absorb, and so the possibility for interference is greater than MS/MS, and more care needs to be taken to prepare clean extracts and validate the method. For this reason, limit of detection (LOD) values are at least an order of magnitude higher than those reported for MS methods [2].

3.8 Sample pretreatment

In many analyses, effective sample preparation is required prior to the determination of melamine and its analogues due to the complexity of the samples [4, 7]. The main objective of sample treatment, including extraction, preconcentration and derivatisation procedures, is to avoid potential interference by coextracted sample components to achieve methods with lower LOD [2, 12]. The strategies of sample pretreatment depend on the analyte, the different matrices and the selective abilities of detection methods [1].

Melamine is a weakly alkaline compound that can hydrolyse in strongly acidic or alkaline solutions. Initial melamine extraction can be carried out under neutral, acidic or alkali conditions. Neutral extraction can be carried out using acetonitrile–water or methanol–water mixtures, while acidic extraction can be carried out using hydrochloroacetic, formic, acetic, hydrochloric or trifluoroacetic acids. Due to the recent worldwide acetonitrile shortage, methanol has been an effective long-term alternative solvent for the extraction and analysis of melamine and its analogues in foods [12].

Aside from centrifugation and filtration, solid-phase extraction (SPE) is most often incorporated into methods for further sample clean-up after extraction. Cation-exchange/reversed phases are used in the analysis of melamine, while anion-exchange sorbents are used to isolate cyanuric acid [2]. As melamine electrolyses to a positive ion in an acidic solution, cation-exchange SPE is often used to remove neutral and acidic interference [12]. Other solid-phase sorbents used for melamine extraction include a molecularly imprinted polymer (MIP), which is a good example of highly selective extraction due to its molecular recognition functions [1, 2, 12]. The MIP is synthesised in the presence of functional monomers, template molecules and a crosslinking agent using covalent, noncovalent and sacrificial spacer methods. MIP has gained wide acceptance due to its predetermined selectivity for target molecules, high affinity, robustness of recognition, stability, low cost and easy preparation [3, 12].

3.9 Issues during analysis

Various issues are encountered during the analysis of melamine and related compounds including contamination, matrix effects and analyte instability, and their severity may vary depending on the method used, the type of matrix and the analytes examined.

In the high-performance liquid chromatography (HPLC)–MS/MS system, persistent background contamination with melamine in the blanks has been caused by carryover. This phenomenon was also noted during GC–MS analysis with cyanuric acid. To minimise the exposure of the system to the target compounds and prevent carryover, only low concentration standards should be injected before the sample. In order to avoid cross-contamination of reusable plastic labware, which could be another possible source of contamination, glassware should be rinsed with an acetonitrile/water solution. One report mentions filter papers that could contain melamine, and so it would be prudent to avoid its use.

The type of matrix can affect the extent to which melamine and related analytes are recovered from a sample during preparation; it can also affect analyte chromatography and ionisation during MS analysis. This is most often noted while using the sensitive HPLC–MS/MS methods. Matrix effects were also observed during GC–MS analysis. This effect is known to vary between instruments, inlet liners and packaging materials, highlighting the need for all methods to be validated prior to use . Matrix components can also affect the derivatisation of melamine and related compounds due to the possible reaction between the matrix components and the derivatising agent. A solution to the problem of matrix effects includes the use of internal standards or matrix-matched standards. To avoid the uncertainties associated with selecting a proper blank matrix for standards, stable isotope-labelled internal standards can be used.

With regard to the stability of melamine and related compounds, in basic conditions, melamine, ammeline and ammelide hydrolyse to cyanuric acid. Aqueous extracts of melamine and cyanuric acid can be stored at 5–10 °C for up to 1 month for further dilution and analysis [2].

3.10 Liquid chromatography methods

In the majority of laboratories, melamine and its metabolites are routinely separated by LC due to the small and polar nature of the compounds, and detected using selective techniques [1, 2]. HPLC is one of the most powerful techniques due to its high efficiency, good reproducibility and simultaneous detection of the presence of melamine and its analogues [3]. Some liquid chromatographic methods for the analysis of melamine and related compounds are listed in Table 3.2.

Table 3.2: LC methods [9, 16–18].

Analytes	Matrix	Detector	Sample preparation or extraction	Column/mobile phase	LOD/(LOQ)	Results	Reference
Melamine, ammeline, ammelide	Eighteen different articles of new melamine plastic kitchenware	DAD	Food simulant acetic acid 3% (w/v) 2 h at 70 °C	Column: Luna HILIC Phenomenex 200 Å (150 mm × 3.0 mm × 3 µm) Isocratic conditions Mobile phase: ammonium formate 10 mM (fitted to pH 3 with formic acid)/acetonitrile (5:95, v/v)	Melamine: 0.1 mg/L (0.25 mg/L) Ammelide: 2 mg/L (4 mg/L) Ammeline: 1 mg/L (2 mg/L)	Melamine migration was detected in eight of the 18 food contact articles (none of the sample exceeded the specific migration limit of 2.5 mg/kg). Ammeline and ammelide were not detected in any sample.	[9]

Melamine	246 samples of 41 types of retail melamine-ware products	ESI–MS/MS	Two types of simulants (3% acetic acid and distilled water) at three test conditions (25, 70 and 100 °C) for 30 min	Column: Altima HP HILIC C18 (150 mm × 2.1 mm × 3 μm) Gradient mode Mobile phase: ammonium acetate/formic acid (0.05%) in water (A) and ammonium acetate/formic acid (0.05%) in acetonitrile (B) (95:5, v/v) The gradient program began at 0.1 min for pump B (80%), the decreases at 5%, hold for 2 min, and finally returning to 80% at 2.01 min and hold until 6 min	LOQ: 5 ng/mL	Melamine migration was detected from all samples. Distilled water: 22.2 ng/mL (25 °C), 49.3 ng/mL (70 °C) and 84.9 ng/mL (100 °C). Acetic acid: 31.5 ng/mL (25 °C), 81.5 ng/mL (70 °C) and 122.0 ng/mL (100 °C)	[16]
Melamine	Melamine leached from tableware	Fluorescence detection (FLD)	First protocol: boiled water or milk Second protocol: water or milk at room temperature and then concentrated Both protocols included derivatizing with 10-methyl-acridone-2-sulfonyl chloride	Column: SB C18 (50 mm × 2.1 mm × 1.8 μm) Gradient mode Mobile phase: 5% acetonitrile in water (A) and acetonitrile (B) Gradient condition of mobile phase was as follows: 20–35% B from 0 to 5 min; 35–100% B from 5 to 8 min	First way: 0.1 μg/L for water and 0.4 μg/L for milk samples (–) Second way: 0.005 μg/L due to the 20 times concentration (–)	Melamine was found in all the samples analysed by the first approach (4.24–428 μg/L), while no melamine as found in the second approach.	[17]

(continued)

Table 3.2 (continued)

Analytes	Matrix	Detector	Sample preparation or extraction	Column/mobile phase	LOD/(LOQ)	Results	Reference
Melamine	13 samples of polypropylene, 3 samples of polycarbonate, 6 samples of melamine resin and 15 dairy product packages	Photodiode array (PDA)	Simulating solutions (*n*-hexane, distilled water, 15% ethanol and 3% acetic acid) with a contact time of 2 h at 60 °C	Column: Kromasil KR100-5 C8 (250 mm × 4.6 mm × 5 μm) Isocratic conditions Mobile phase: buffer (10 mmol/L citric acid and 10 mmol/L sodium octane sulfonate): acetonitrile (85:15, v/v)	0.18 mg/L (0.18 mg/kg)	Melamine was confirmed in 3 of 6 melamine containers. The migration in 15% ethanol and 3% acetic acid was greater than in distiller water.	[18]

(–): not available
ESI: Electrospray ionisation
FLD: Fluorescence detection
HILIC: Hydrophilic-interaction chromatography
LOQ: Limit of quantification

As the UV spectra for melamine exhibits absorption bands below 250 nm, erroneous quantification can occur if insufficient care is paid to chromatographic conditions or if sample preparation is not optimised. To increase sensitivity and confidence in the reported results, several methods were proposed, including DAD systems and FLD [12].

MS techniques are widely used to determine contaminants due to their molecular specificity and high sensitivity of detection. Both LC–MS and LC–MS/MS are currently the main methods for the quantitative determination of melamine [12]. Melamine is analysed in positive ESI and cyanuric acid in negative ESI when using HPLC–MS/MS. The opposite polarities required necessitate the use of polarity switching during one HPLC–MS/MS run or two different HPLC–MS/MS runs. Monitoring at least two transitions is recommended for MS analysis, as the ratios of responses from these two transitions can be calculated and used as another identification criterion, and to ensure accurate identification of coeluting interferences during the analysis of melamine. The most common transitions monitored for melamine are 127→85 and 127→68, which correspond to a loss of cyanamide from the $[M+H]^+$ ion to produce the 2,4-diamino-1,3-diazete cation, and subsequent loss of ammonia to form the ion with *m/z* 68. The fragmentation of $[M-H]^-$ for cyanuric acid is similar to the pathways proposed for melamine. Atmospheric pressure chemical ionisation (APCI) in a positive ionisation mode has also been used for melamine, and APCI in a negative ionisation mode for cyanuric acid, ammeline and ammelide. The APCI method using a heated nebuliser probe was found to be more sensitive than ESI for several of the compounds [2].

Melamine as a polar molecule is a good candidate for normal-phase LC, HILIC, where a polar hydrophilic analyte partitions between a relatively polar stationary phase and a relatively nonpolar mobile phase [4]. However, poor retention and/or separation of the small and very polar melamine and coextracted matrix components on traditional reversed-phase LC columns (e.g., the traditional C_{18} or C_8 column) has been noted [2].

Many methods utilise ion-exchange, ion-pair chromatography or HILIC [1]. HILIC columns are designed for the retention and separation of polar compounds, and so it has been successfully applied to separate melamine from analogous compounds and interfering materials [12]. Ion-pair reagents facilitate separation on a reversed-phase column, but may result in ionisation suppression in the MS source or decreased UV absorption. It was shown that multiple columns in series improve the separation of melamine and analogues [2].

3.11 Gas chromatography methods

GC is another common method used to confirm and quantify melamine and related compounds.

However, this technique usually needs an extra pretreatment phase of derivatisation with three methyl silanised reagents due to the polar nature of melamine and related compounds, and is therefore relatively nonvolatile [1, 2]. The trimethylsilyl derivates are analysed in the positive electron impact (EI) mode in GC–MS/MS. Single-stage GC–MS methods also use EI selected ion monitoring of multiple ions to quantify the trimethylsilyl derivates. Single-stage MS methods are less selective than MS/MS methods, which may result in a lowered sensitivity [2].

However, the instrumentation is expensive and sample pretreatment is time consuming, and so the methodology is impractical and inefficient for analysing a large number of samples [3, 12].

3.12 Other techniques

Alternatives to costly chromatographic methods include cheaper screening methods with sufficient sensitivity and selectivity. These methods often provide results more quickly than the more selective methods, but they lack the power to unequivocally identify melamine and related compounds as do the MS-based methods [2].

3.12.1 Enzyme-linked immunosorbent assay

All immunoassays are based on the highly selective antigen–antibody interaction. As melamine has a small MW, it is difficult to stimulate the immune response and subsequently produce a specific antibody [12]. There is no information readily available regarding the applicability of enzyme-linked immunosorbent assay (ELISA) methods for the analysis of the analogues of melamine, i.e., cyanuric acid, ammeline, ammelide or cyromazine [2].

ELISA is a good screening method that can be used for a large number of samples because it has the advantages of simplicity, low cost, rapid turnaround time and high throughput. However, the procedure requires labour-intensive operations (incubation, washing and enzymatic reactions); it is not able to detect concentrations in sample matrices as low as MS/MS methods, and there is the possibility of false-positives as most of the antibodies show cross reactivity to the structure of related compounds of melamine [1–3]. So, when a positive sample is found during screening, it must be rechecked by a more selective method to identify the compound detected [2, 4].

Commercial ELISA test kits are also developed for quantitative analyses. Using the ELISA test kit, the melamine from samples and melamine horseradish peroxidase conjugate were bound to the melamine antibody. After washing, a clear substrate was added and any bound enzyme conjugate resulted in a blue colour, which was compared with standards to determine the melamine concentration in the sample [4].

3.12.2 Vibrational spectroscopy

Vibrational spectroscopies such as surface-enhanced Raman spectrometry (SERS), near infrared (NIR) and midinfrared spectroscopy (MIR) have been used to detect melamine in various food matrices [12].

NIR and MIR spectroscopic methods are rapid, nondestructive and require little or no sample pretreatment, although the preparation of calibration standards in the appropriate matrix is required as determination is matrix dependent [2].

Raman spectrometry is a rapid and cost-effective method for the detection of melamine with minimum sample preparation [1]. SERS is a type of Raman spectrometry that measures molecular vibrations by light scattering [19]. He and co-workers (2008) developed a method using SERS coupled with gold nanosubstrates that can quickly detect and characterise a small amount of melamine and its analogues in aqueous solutions. However, the commercial substrates are costly and the background noise from substrates increases over time [12, 19].

3.12.3 Biosensors

New advances in chemistry, physics and biology have led to the development of biosensors that can detect a large range of biological elements with selectivity and sensitivity.

An electrochemical biosensor is composed of a biosensor and an electrochemical transducer, and is considered to be a chemically modified electrode as an electronic conducting, semiconducting or ionic conducting material is coated with a biochemical film. It is a viable alternative for the detection of melamine due to its high sensitivity, selectivity, low cost, fast response, simplicity and possibility for miniaturisation, portability and integration into automated devices. Moreover, the biosensor only requires a small amount of sample [3, 20].

Several colorimetric systems have been designed to detect melamine using nanoparticles, which have unique properties such as biostability, biofunctionalisation and spectral [3, 12]. The presence of melamine induces the aggregation of unmodified gold nanoparticles (AuNP) and a consequent visible colour change. AuNP bind to melamine based on the electrostatic reaction between the negatively charged gold surface and the positively charged amino groups of melamine [3, 20].

3.13 Quantitation

The presence of observed matrix effects prompted the use of various quantitation systems such as matrix-matched standards and stable isotope-labelled internal standards to compensate for these effects during quantitation and data analysis.

Stable isotope-labelled internal standards are useful for MS-based methods because their physico-chemical properties are as close as possible to the analytes to be measured. Only $^{13}C_3$-melamine, $^{13}C_3$-cyanuric acid, $^{15}N_3$-melamine, $^{15}N_3$, $^{13}C_3$-melamine and $^{15}N_3$, $^{13}C_3$-cyanuric acid are currently available as internal standards. When they are added prior to sample extraction, they can account for a loss of analyte during sample processing and volume changes; and when they are added after sample preparation, but before MS analysis, they can avoid changes during ionisation in the MS source [2].

The preparation of labelled internal standards $^{13}C_3$-melamine and $^{13}C_3$-cyanuric acid using the common substrate $^{13}C_3$-cyanuric chloride was developed by Varelis and Jeskelis in 2008 [21].

References

[1] Y. Liu, E.E. Todd, Q. Zhang, J.R. Shi and X.J. Liu, *Journal of Zheijiang University-Science B (Biomedicine and Biotechnology)*, 2012, 13, 7, 525.
[2] S.A. Tittlemier, *Food Additives & Contaminants*, 2010, 27, 2, 129.
[3] K. Rovina and S. Siddiquee, *Journal of Food Composition and Analysis*, 2015, 43, 25.
[4] Y.C. Tyan, M.H. Yang, S.B. Jong, C.K. Wang and J. Shiea, *Analytical and Bioanalytical Chemistry*, 2009, 395, 3, 729.
[5] J.R. Ingelfinger, *England Journal of Medicine*, 2008, 359, 26, 2745.
[6] *Background Paper on the Chemistry of Melamine Alone and in Combination with Related Compounds*, World Health Organization, 2008. http://www.who.int/foodsafety/fs_management/Melamine_2.pdf.
[7] K. Sharma and M. Paradakar, *Food Security*, 2010, 2, 1, 97.
[8] J.G. Wen, X.J. Liu, Z.M. Wang, T.F. Li and M.L. Wahlqvist, *Asia Pacific Journal of Clinical Nutrition*, 2016, 25, 4, 697.
[9] V.G. Ibarra, A. Rodríguez Bernaldo de Quirós and R. Sendón, *European Food Research and Technology*, 2016, 242, 8, 1187.
[10] B. Kim, L.B. Perkins and R.J. Bushway, *Journal of AOAC International*, 2008, 91, 2, 408.
[11] C.B. Stine, R. Reimschuessel, Z. Keltner, C.B. Nochetto, T. Black, N. Olejnik, M. Scott, O. Bandele, S.M. Nemser, A. Tkachenko, E.R. Evans, T.C. Crosby, O. Ceric, M. Ferguson, B.J. Yakes and R. Sprando, *Food and Chemical Toxicology*, 2014, 68, 142.
[12] F. Sun, W. Ma, L. Xu, Y. Zhu, L. Liu, C. Peng, L. Wang, H. Kuang and C. Xu, *Trends in Analytical Chemistry*, 2010, 29, 11, 1239.
[13] N. Rai, D. Banerjee and R. Bhattacharyya, *Nutrition*, 2014, 30, 4, 380.
[14] European Food Safety Authority in Scientific Opinion on Melamine in Food and Feed, *EFSA Journal*, 2010, 8, 4, 1573.
[15] A.J. Dane and R.B. Cody, Analyst, 2010, 135, 696.

[16] Z. Chik, D.E. Mohamad Haron, E.D. Ahmad, H. Taha and A.M. Mustafa, *Food Additives & Contaminants: Part A*, 2011, 28, 7, 967.
[17] S. Zhang, Z. Yu, N. Hu, Y. Sun, Y. Suo and J. You, *Food Control*, 2014, 39, 25.
[18] J. Lu, J. Xiao, D.J. Yang, Z.T. Wang, D.G. Jiang, C.R. Fang and J. Yang, *Biomedical and Environmental Sciences*, 2009, 22, 2, 104.
[19] L. He, Y. Liu, M. Lin, J. Awika, D.R. Ledoux, H. Li and A. Mustapha, *Sensing and Instrumentation for Food Quality and Safety*, 2008, 2, 1, 66.
[20] N. Chen, Y. Cheng, C. Li, C. Zhang, K. Zhao and Y. Xian, *Microchimica Acta*, 2015, 182, 11–12, 1967.
[21] P. Varelis and R. Jeskelis, *Food Additives & Contaminants: Part A*, 2008, 25, 10, 1208.

4 Phthalates and alternative plasticisers

4.1 Phthalates

Plasticisers are one of the most common additives used in wide-ranging applications, such as polymer processing. The desirable properties of plasticisers include increasing the flexibility of materials and workability of complex formulations; improving the impact resistance of materials; reducing the viscosity of polymers in order to improve film elasticity and tensile strength; improving the low-temperature properties of materials; increasing elongation and decreasing tensile strength; reducing the pressure of extraction and improving the processing and blending of polymers [1]. The most widely used plasticisers in the world are esters of phthalic acid, which combine most of the desirable properties of plasticisers.

Phthalates (PAE) are a group of chemical compounds, also called phthalic acid esters, that are synthesised via esterification between phthalic anhydride and oxo alcohols [ranging from C_4 to C_{13} alcohols for polyvinyl chloride (PVC) applications] [2, 3]. These compounds are widely used in a variety of industrial applications, including cosmetics, personal care products, detergents, pesticides, adhesives, paints, inks, sealants, medical devices, plastics [food contact materials (FCMs)] and toys. The industrial application of PAE mainly depends on the physical properties of these compounds [4–6].

In general, PAE can be distinguished by their molecular weight (MW); low MW PAE have short alkyl side chains with one, two or four carbon atoms and high MW PAE have long alkyl side chains with four to 10 carbon atoms. Low MW PAE, such as dibutyl phthalate (DBP) and diethyl phthalate (DEP), are not used as the sole plasticiser in cellulosic, urethane and acrylic resins. These PAE are most frequently used in cosmetic applications as solvents in personal care products and in lacquers, adhesives, printing inks, varnishes and coatings [5, 7].

Due to their properties, that is, high solubility in polymeric materials, stability, low volatility and fluidity, high MW PAE, such as di-*n*-octyl phthalate (DnOP) and diisononyl phthalate (DiNP) are used mainly as plasticisers in the manufacture of different materials, giving them greater flexibility and enhancing their technical properties [8, 9]. The main application of high MW PAE is in the production of PVC, although they are also used in other materials such as polyurethane, polyvinyl acetate, polyvinylidene chloride and cellulosic materials [7, 8]. An overview of the most important PAE, including their physicochemical properties, is given in Table 4.1.

PAE are also used as solvents and/or plasticisers in paints, inks and adhesives. Although printing inks are not in direct contact with food, it has been found that the plasticisers in them can migrate to the food through the packaging material, or during storage in reels, via the 'set-off' effect. Other PAE were found to originate from adhesives used in the joints of the packaging.

https://doi.org/10.1515/9783110648065-004

Table 4.1: Structures and physicochemical properties of PAEs.

Structure	Common name	Acronym	CAS no.	Molecular formula	MW	MP (°C)	BP (°C)	Log *P* (o/w)	Vapour pressure (mm Hg)	SML[a] (mg/kg)
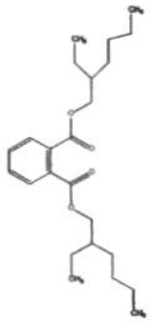	Bis (2-ethylhexyl) phthalate	DEHP	117-81-7	$C_{24}H_{38}O_4$	390.56	−55*	230*	7.6	1.42E-07	1.5 60**
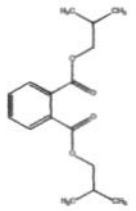	Diisobutyl phthalate	DIBP	84-69-5	$C_{16}H_{22}O_4$	278.34	−64*	296	4.11	6.65E-03	–
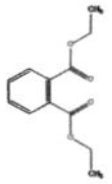	Diethyl phthalate	DEP	84-66-2	$C_{12}H_{14}O_4$	222.24	−40.5	295	2.42	0.0021	–
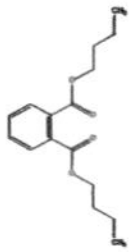	Dibutyl phthalate	DBP	84-74-2	$C_{16}H_{22}O_4$	278.34	−35	340	4.5	2.01E-05	0.3 60**

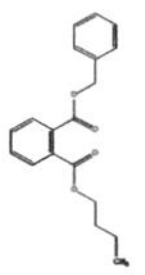	Butyl benzyl phthalate	BBP	85-68-7	$C_{19}H_{20}O_4$	312.36	−35*	370	4.73	8.25E-06	60
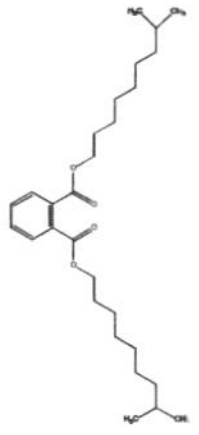	Diisodecyl phthalate	DIDP	68515-49-1 26761-40-0	$C_{28}H_{46}O_4$	446.66	−45.6	250−257 *	10.36	5.28E-07	9 60**
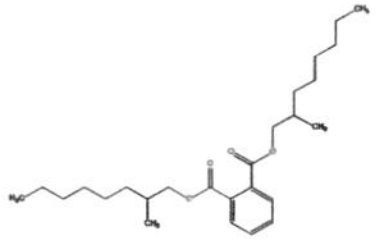	Diisononyl phthalate	DiNP	28553-12-0 68515-48-0	$C_{26}H_{42}O_4$	418.61	−43*	244−252 *	9.370	5.40E-07	9 60**
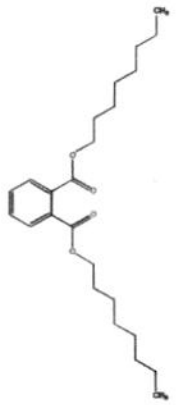	Di-*n*-octyl phthalate	DnOP	117-84-0	$C_{24}H_{38}O_4$	390.56	−25*	220	8.1	1.00E-07	–

(continued)

Table 4.1 (continued)

Structure	Common name	Acronym	CAS no.	Molecular formula	MW	MP (°C)	BP (°C)	Log *P* (o/w)	Vapour pressure (mm Hg)	SML[a] (mg/kg)
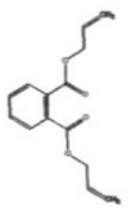	Diallyl phthalate	DAP	131-17-9	$C_{14}H_{14}O_4$	246.26	−70*	158–165*	3.23	$1.6E^{-03}$	ND
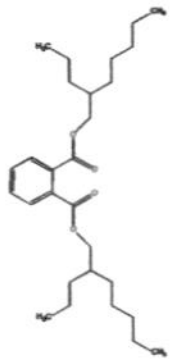	Di(2-propylheptyl) phthalate	DPHP	53306-54-0	$C_{28}H_{46}O_4$	446.66	NA	425.8**	10.554*	1.86E-7*	

Note: Data obtained from Chemidplus version web.
* Experimental data obtained from SciFinder® version web; N/A, not available; ** SML(T) (group restriction).
[a] Commission regulation (EU) No. 10/2011.

Not all plastics contain plasticisers, for example, polyethylene (PE) and polypropylene. However, PAE has been found in water bottles made from polyethylene terephthalate (PET), PE and glass. Although these materials do not contain PAE, contamination in water bottles can occur during bottling processes or from the cap liner [10].

Some of the most widely used PAE in diverse applications are DEHP, DBP, DiNP, DiDP (diisodecyl phthalate), BBP, DIBP, DEP and DnOP. The main applications of the mentioned PAE are summarised in Table 4.2.

It has been estimated that approximately 51% of the global production of plasticisers is DEHP, being the predominantly used plasticiser in several applications [6].

Since their introduction in 1930, DEHP has been the most widely used plasticiser in the production of flexible PVC products. DEHP is present in a great variety of consumer products and hence DEHP exposure is widespread in the population. In 2011, PAE drew the specific attention of the general public due to a food safety incident in Taiwan, which was related to the illegal addition of DEHP and DiNP as clouding agents in food and beverages. The ingredient commonly used as a clouding agent in Taiwan is palm oil. Food products related to the incident included sports drinks, concentrated juice beverages, tea drinks and food supplements in capsule or powder form [26].

As a result, in recent years there has been increasing interest in monitoring PAE compounds in a broad range of matrices, and several analytical methods for measuring the amount of PAE in various food simulants and matrices have been developed and published.

PAE may be found in food products as a result of its migration from FCM; however, it is not the only source of PAE contamination. PAE may also be present in foods as a consequence of environmental contamination. There are several routes of human exposure to PAE, including oral, dermal and inhalation.

Exposure also includes medical procedures (blood transfusions, long-term haemodialysis). However, it has been reported that human exposure to PAE occurs mainly through food ingestion [6].

Several studies have shown the potential of some PAE to be endocrine disrupting chemicals, with the degree of toxicity depending upon the molecular nature. PAE are rapidly metabolised and excreted in urine and faeces. Human exposure to PAE can be routinely monitored through the measurement of urinary concentrations of PAE metabolites, which have also been detected in saliva and amniotic fluid [13]. Table 4.2 lists the common metabolites for the biomonitoring of certain PAE.

Several animal studies have shown a variety of toxic effects as a result of PAE exposure, which include liver and kidney toxicity, carcinogenicity and damage to the reproductive system, among others. Genotoxicity studies suggest an associated damaging impact of DBP and DIBP on deoxyribonucleic acid in human mucosal cells and lymphocytes, respectively [27].

Table 4.2: Principal function and applications of PAEs.

Compound	Function	Products	Metabolite abbreviation	Reference
DEHP	Plasticiser in PVC Other polymers where DEHP are used are other vinyl resins and cellulose ester plastics such as cellulose nitrate, cellulose acetate butyrate, vinyl chloride/vinyl acetate copolymer with cellulose acetate butyrate, ethyl cellulose, polymethyl methacrylate, polystyrene, vinylchloride/vinyl acetate copolymer.	PVC applications: Tablecloths, shower curtains, furniture and automobile upholstery, imitation leather, garden hoses, floor tiles, swimming pool liners, sheathing for wire and cable, rainwear, shoes, toys, dolls, food packaging materials, medical devices (blood and intravenous bags, catheters, tubing for dialysis, oxygen masks), surgical gloves. Non-polymer applications: Adhesives, lacquers and paints, ceramics, sealants, cosmetics, inert ingredient in pesticides, dielectric fluids, among others.	Mono-(2-ethyl-5-hydroxyhexyl)phthalate (MEHHP) Mono-(2-ethyl-5-oxohexyl) phthalate (MEOHP) Mono-2-ethylhexyl phthalate (MEHP) Mono-2-ethyl-5-carboxypentyl phthalate (MECPP)	[11–14]
DIBP	Plasticiser for nitrocellulose, cellulose ether and polyacrylate and polyacetate dispersions. Is too volatile for use in PVC; it is often combined with other PAEs.	Printing inks for paper and packaging nail polish, cosmetics, personal care products lubricants, rubber dentistry settings, as a fuel stabiliser, in leather varnishes and lacquers, explosive material, lacquer manufacturing and methyl methacrylate applications.	Mono-isobutyl phthalate (MiBP) 2OH-mono-isobutyl phthalate (2OH-MiBP) 3OH-mono-isobutyl phthalate (3OH-MiBP)	[13–15]

DEP	Used in the production of cellulose esters. Blends well (up to 70%) with cellulose nitrate, polyacrylates and polymethyl methacrylates. Extensively used as a denaturant for cosmetics.	Plastic packaging films, cosmetic formulations, personal care products, toiletries, detergents and insecticides, medical treatment tubing, as a dye application agent, adhesive and sealant	Monoethyl phthalate (MEP)	[13,16,17]
DBP	Used as a plasticiser for polyvinyl acetate, nitrocellulose and PVC. Plasticiser in adhesives, paints, lacquers.	Used as plasticiser in resins, polymers, adhesives, paints, lacquers. Printing inks, sealants/grouting agents, medical supplies, pharmaceutical coatings, insecticides, nitrocellulose paints, film coatings and glass fibres, cosmetics (perfume solvent, lubricant for aerosol, skin emolient, plasticiser in nail polish, among others)	Mono-isobutyl phthalate (MiBP)	[13,18,19]
BBP	Mainly used as plasticiser in polymeric materials. Compatible with PVC, PVA, polyacrylates, nitrocellulose. Plasticiser in the production of vinyl floor tiles and adhesives, in food containers and wrapping materials. Other applications in plastics include polymethyl methacrylates (PMMA), polyamide, and thermoplastic polyester	Adhesives (construction, paper), coatings and inks (car care products, construction, paper, board), sealants, food wrap, food packaging materials, traffic cones, among others.	Mono-*n*-butyl phthalate (MBP) Mono-benzyl phthalate (MBzP)	[13,16,20]

(continued)

Table 4.2 (continued)

Compound	Function	Products	Metabolite abbreviation	Reference
DIDP	It is mainly used as a plasticiser for PVC applications, as well as in polyvinyl acetates, celluloses and polyurethanes.	Wire and cable, flooring, food contact plastics, toys, film and sheet, building and construction, among others. Non-PVC applications: include sealants, adhesives, rubbers, lubricants, paints and lacquers.	Mono(carboxy-isononyl) phthalate (MCiNP) Mono(hydroxy-isodecyl) phthalate (MHiDP) Mono(oxo-isodecyl) phthalate (MOiDP)	[6,21,22]
DiNP	Plasticiser in PVC (approximately 95%)	Plasticiser in PVC applications (approximately 95%) Non-PVC applications: inks and pigments, adhesives, sealants, paints and lacquers. electrical cables, packaging, stationery, flooring and toys, inks and pigments, adhesives, sealants, paints and lacquers and lubricants	Mono-iso-nonyl phthalate (MiNP)	[13,23,24]
DnOP	Primarily used as a plasticiser in the production of PVC and copolymers. Also used as plasticiser for polystyrene, cellulose nitrate, among others.	Automobile industry (traffic cones, automobile hoses), home maintenance (flooring tiles, carpet tiles, swimming pool liners, insulations, floor adhesives, etc.), inside home (shower curtains, wall coverings, tablecloths), cosmetics, pesticides, children toys, shoes, packaging films, among others.	Mono-(3-carboxypropyl) phthalate (MCPP)	[13,16,25]

PAE have also been associated with adverse effects on the genitals of male infants through prenatal exposure (reduction in the anogenital distance). DIBP exposure in pregnant rats showed evidence of embryotoxicity and teratogenicity [28].

PAE such as DEHP have been associated with reproductive toxicity and carcinogenicity. DEHP mainly affect the male reproductive system through its main metabolite, MEHP. Similarly, it has been reported that DEHP can induce hepatic tumours in mice and rats [29].

Based on toxicological studies, the European Food Safety Authority has established the tolerable daily intake (TDI) for PAE at 0.01 mg/kg bodyweight per day for DBP, 0.5 mg/kg bodyweight per day for BBP, 0.05 mg/kg bodyweight per day for DEHP and the sum of DiNP and DiDP (0.15 mg/kg bodyweight) [30–34].

4.1.1 Determination of phthalates in food

Analysis of PAE in food matrices generally requires extraction and clean-up procedures prior to separation and quantification. Samples have to be homogenised prior to extraction, which can be achieved via shaking, stirring or mixing; however, liquids do not generally require any homogenisation treatment.

The literature details a variety of extraction methods for PAE analysis, which include liquid–liquid extraction (LLE), solid-phase extraction (SPE), solid-phase microextraction (SPME), molecularly imprinted solid-phase extraction (MISPE), magnetic solid-phase extraction (MSPE) and Quick, Easy, Cheap, Effective, Rugged and Safe (QuEChERS), among others [35–39].

LLE or solvent extraction consists of separating the compounds based on their solubility in two different immiscible liquids. For this purpose, an adequate amount of solvent is necessary to capture all analytes from the original sample. Several authors have used this method for PAE extraction in food products. Zaater et al. [40] employed LLE with a mixture of methylene chloride–petroleum ether (20:80, v/v) for the isolation and enrichment of six PAE [dimethyl phthalate (DMP), DEP, DBP, BBP, DEHP and DnOP] without further clean-up. Gas chromatography (GC)–mass spectrometry (MS) was used for identification and high-performance liquid chromatography (HPLC) with ultraviolet (UV) detection for quantification. In 2015, Milojković et al. [41] evaluated 14 LLE techniques for the determination of four PAE (DMP, DBP, BBP and DEHP) in milk-based samples. The study showed that *n*-hexane was the best extractant, obtaining a homogeneous and clear extract with a precipitated raffinate phase. Higher recovery is observed for the high MW PAE (log Kow = 7.6 for DEHP), while low MW PAE are more water soluble, relatively hydrophilic and thus the recovery decreases (log Kow = 1.5 for DMP).

Other authors employed the liquid-phase microextraction (LPME) method for PAE determination in food samples. This promising low-cost technique emerged from LLE, and combines extraction, concentration and sample introduction in one step, and allows analyte extraction in a few microlitres of organic solvent [42]. Farahani et al. [43] developed an LPME technique using a directly suspended organic microdrop coupled with GC–MS for the extraction and determination of DMP, DEP, DAP, di-*n*-butyl phthalate (DnBP), BBP, dicyclohexyl phthalate (DCHP) and DEHP in water samples. Some efficiency factors for microextraction, such as temperature, organic solvent, salt effect, stirring rate and extraction time, were optimised in this study. Several different modes of LPME have been developed, among the main ones it is worth mentioning dispersive liquid–liquid microextraction (DLLME), single-drop microextraction [44] and hollow fibre-LPME [45].

Farajzadeh et al. [46] used DLLME for the extraction, preconcentration and determination of five PAE (DMP, DEP, DIBP, DBP and DEHP) in cow's milk samples. Acetonitrile (ACN) was used as the extraction solvent and 1,2-dibromoethane was selected as the extraction solvent in the DLLME step. After the extraction, the selected compounds were determined by GC–flame ionisation detection (FID) and GC–MS.

DLLME emerged as a powerful preconcentration method, which is based on a ternary solvent system such as homogeneous LLE and cloud-point extraction [47]. For this procedure, the selection of an appropriate organic extraction solvent is a very important factor. Recently, Yan et al. [48] developed a method for DMP, DEP, DBP, BBP, diisooctyl phthalate (DIOP) and DnOP analysis in bottled milk using ultrasound-assisted DLLME followed by GC–FID. In this study, methanol was selected as the dispersant due to its higher dispersing capability for the extractant and relatively decreased analyte loss.

SPE is another useful technique for rapid and selective sample preparation. Most commonly, SPE is achieved using miniature column or cartridge devices. Compared with LLE, SPE requires lower utilisation of organic solvents and can be used as an effective clean-up method to remove matrix interferences and improve sensitivity by preconcentration of the analytes [49]. Mohamed and Ammar [50] developed an SPE method followed by capillary GC coupled to MS for the determination of five PAE (DMP, DEP, DBP, BBP and DEHP) and bis(2-ethylhexyl) adipate (DEHA) in traditional Egyptian food and drink. High accuracy and good recovery of the analytes were obtained in this study. In the same way, Liu et al. [51] used an SPE method before GC–tandem mass spectrometry (MS/MS) analysis for the determination of 18 PAE in edible vegetable oils. A silica/PAS column was used for purification of the sample (supernatant), which required previous ultrasonic extraction with ACN.

The SPME technique has been selected by several authors for the determination of analytes at trace levels. This technique has been available for sample preparation since 1993, and has several advantages including relatively low cost, simplicity of the procedure, elimination of organic solvents from the protocol and compatibility with HPLC, GC and capillary electrophoresis, among others. Extraction is performed via immersion (fibre in the medium under investigation) and headspace (above the medium under investigation) modes [52, 53]

Chen et al. [54] employed headspace SPME, in conjunction with GC, using a calyx[6]arene fibre for the determination of six PAE [DBP, diamyl phthalate (DAmP), DnOP, DEHP, dinonyl phthlate (DNP) and didecyl phthlate (DDP)] in non-alcoholic beverages. The results indicated that the extraction temperature was the most significant factor for all PAE in the headspace SPME procedure.

Other authors have used a new mode of SPE termed MSPE, for example, Wang et al. [39] synthesised magnetic Fe_3O_4 microspheres (Fe_3O_4@SiO_2-G) via a chemical bonding method, and applied it for the extraction of DAP, dipropyl phthalate (DPRP), BBP, DCHP and DEHP in soybean milk prior to HPLC analysis. In this study, several parameters that could affect the extraction efficiency were investigated (amount of Fe_3O_4@SiO_2-G, extraction time, pH of sample solution and the desorption conditions).

MSPE is an extraction method that is based on the use of magnetic or magnetisable adsorbents. This method combines the advantage of magnetic separation technology and SPE, and makes sampling and collection easier and faster. The separation process can be performed directly for samples containing suspended solid materials without the need for additional centrifugation or filtration [39, 55, 56].

Due to the insufficient selectivity of some sorbents for SPE, such as inorganic oxides and low specific sorbents, over the last few decades, a growing trend has been focused on class-specific sorbents such as molecularly imprinted polymers (MIPs). MIPs have become an interesting research field due to their advantages and low cost of preparation, as well as their potential application to a wide range of target molecules. MIPs are artificial polymers with specific binding sites for a particular analyte [57]. MIPs are employed as sorbents in the MISPE technique.

He et al. [58] developed an MISPE technique for DBP extraction from soybean milk. In this report, DBP was employed as the template molecule and MIPs were synthesised via the bulk polymerisation of methacrylic acid.

QuEChERS is another procedure that has become very popular over the last few years for the determination of different compounds in food matrices. This technique uses a single extraction in ACN, followed by liquid–liquid partitioning after the addition of a mixture of anhydrous magnesium sulphate ($MgSO_4$) and sodium chloride (NaCl), and a clean-up step in which the extract is mixed with a primary secondary amine (PSA) sorbent and anhydrous $MgSO_4$ [59].

Yin et al. [60] developed a modified QuEChERS method for the GC–MS/MS determination of 16 PAE in tea samples. In order to minimise interference and obtain accurate results, a combination of 200 mg of PSA, 100 mg of graphitised carbon black and 100 mg $MgSO_4$ was applied during the purification process.

Recently, a new microextraction technique has been developed for organic compounds, known as ultrasound-assisted emulsification microextraction. This procedure is a rapid and nonexpensive alternative to other extraction methods and employs a small amount of organic solvents [61].

Yan et al. [62] developed a new ultrasound-assisted surfactant-enhanced emulsification microextraction method using a nonionic surfactant (Triton X-100) as an emulsifier for the extraction and determination of DEP, DMP, DIOP, DBP and DnOP in beverage products. In this study, the determination of trace levels of PAE was achieved without the application of organic dispersive solvents.

Stir bar sorptive extraction (SBSE) is another sample preparation method for the preconcentration of organic compounds into a polydimethylsiloxane-coated stir bar [63]. Cacho et al. [64] developed a method for the determination of PAE and alkylphenols in vegetables using SBSE coupled to GC–MS. Sample extraction was performed using methanol, a solvent capable of quantitatively extracting analytes from vegetable samples. In the same way, higher extraction efficiency for all the analytes was obtained using 10 mL of solvent.

4.1.2 Analytical methods

Several chromatographic techniques have been described for the determination of PAE in food matrices. GC–MS is the most common analytical method for detecting and measuring PAE in food. Since all PAE have similar spectra, it is difficult to identify which PAE is present in a complex sample; however, GC–MS/MS was shown to have several advantages. GC–FID and GC–electron capture detection are alternative techniques.

In general, the majority of methods for the identification and quantification of PAE in food matrices involve GC coupled with MS, MS/MS or FID. However, unequivocal confirmation of PAE can be achieved using MS/MS. Ostrovsky et al. [65] studied a method for the determination of total PAE (DMP, DBP, DEHP, DDP and DEP) in fatty foods using GC–FID, with the limit of detection (LOD) and limit of quantification (LOQ) of 0.4 (2.1) and 1.2 (6.2) μg/g, respectively. The method has proven sufficiently sensitive for the determination of all PAE in real samples.

GC/MS in combination with the sample pretreatment of ACN extraction and a silica/*N*-(*n*-propyl)ethylenediamine-mixed SPE column was used for the determination of 17 PAE in edible vegetable oil. Recoveries ranged from 78.3% to 108.9% for all PAE and the method detection limits were 0.1–0.2 mg/kg [66].

Cavaliere et al. [67] used GC–MS/MS for the unequivocal confirmation and quantification of DMP, DEP, DBP, BBP, DEHP and DnOP at low LOD in fatty matrices (olive oil). Results revealed the presence of DBP, BBP and DEHP at higher concentrations in refined oils and lower levels in extra virgin oils.

HPLC might also be employed in combination with UV detection or MS/MS.

HPLC–UV was used for the determination of DAP, dipentyl phthalate (DPP), DIBP, di-sec-octylphthalate and DnOP in mineral water, juice and milk samples.

The LOD of the method was between 0.2 and 1.2 ng/mL for water and juice samples, and between 2.5 and 5.0 ng/mL for milk samples, and values were comparable with those of other reported methods for the determination of PAE [68].

Mortensen et al. [69] developed an analytical procedure for the determination of six PAE monoesters (monomethyl phthalate, MEP, MBP, MBzP, MEHP and mono-isononyl phthalate) in human milk, consumer milk and infant formula using liquid chromatography (LC)–MS/MS. Detection limits were in the range of 0.01–0.5μg/L. Recovery and repeatability showed that the method is robust and reliable for complex matrices.

Over the last few years, there have been reports of the determination of PAE in wine and dried baby food matrices using GC–ion trap MS [70, 71]. Direct analysis in a real-time ion interface was also shown to be effective in several applications including the analysis of food products and food packaging materials [72, 73]. Guo et al. [74] developed a method for the determination of DBP in wine via flow-injection chemiluminescence analysis. The method was based on the quenching effect of DBP on the luminol–myoglobin chemiluminescence system. Chen et al. [75] studied the determination of DEHP in drinks using membrane filtration enrichment and diffuse reflectance UV spectroscopy.

PAE may be present in food as a result of migration from FCM or via contact with printing inks or adhesives applied to plastic materials that contain PAE. The PAE are not covalently bound to the polymer; therefore, migration of these compounds from packaging into food may occur, especially when they are in contact with fatty foods. However, food contamination by PAE can occur during processing, handling and transportation, or due to its widespread presence as an environmental contaminant, which can be found in water, soil, air and food [76].

Over the last few years, several cases of PAE contamination of food products have been reported. These notifications are reported by the Rapid Alert System for Food and Feed (RASFF). In Europe, the RASFF system allows an effective exchange of information between the European Union member states regarding the risks identified in food, feed or FCM.

Since 2010, 45 notifications involved the migration of PAE from food packaging materials. DEHP is the most widely used PAE in diverse applications including food packaging and was involved in more than half of the notifications (24 notifications). Other PAE involved in the notifications were DiNP (nine notifications), DBP (four notifications), DPHP (two notifications) and DiDP (one notification). In 2015 and 2016, the number of RASFF notifications for PAE from FCM decreased to four, compared with the 2010 and 2011 that registered eight and 14 notifications, respectively [77] (see Table 4.3).

The SML for some PAE and other plasticisers are detailed in Regulation (EU) No. 10/2011 on plastic materials and articles intended to come into contact with food. When there are no SML or other restrictions, a generic SML of 60 mg/kg is applied [78].

Table 4.3: PAEs RASFF notifications since 2010.

	RASFF notifications						
	Year						
Compound	**2010**	**2011**	**2012**	**2013**	**2014**	**2015**	**2016**
DEHP	7	6	3		1	1	2
DiNP	1	4		1			
DEHP + DiNP		1	1				
DBP + DEHP		3	1				
DEHP + bis(2-ethylhexyl) terephthalate (DOTP)			1				
BBP + DEHP				1			
Di(2-propylheptyl) phthalate (DPHP)					1		1
DIDP + DiNP						1	

The migration of PAE is influenced by a number of factors, such as the fat content of the food, time and temperature of contact, the polymer type and concentration of plasticiser in the material, and the length of the storage period [79]. According to the literature reviewed, the PAE most commonly found in several food products was DEHP.

Fierens et al. [80] evaluated the effect of cooking at home on the levels of eight PAE (DMP, DEP, DIBP, DBP, BBP, DEHP, DCHP and DnOP) in various food types (starchy products, vegetables, meat and fish). Food products were analysed before as well as after cooking and, in general, PAE concentrations in food declined after cooking, the exception being in vegetables. In the same way, DEHP was the most abundant compound, before as well as after cooking, followed by DIBP and BBP. In another study by the same authors, the occurrence of DMP, DEP, DIBP, DBP, BBP, DEHP, DCHP and DnOP in 400 food products and packages sold on the Belgian market was explored. DEHP was the most abundant compound (concentrations ranged from not detectable up to 5,932 μg/kg in fish products), followed by DIBP, DBP and BBP [81].

In 2013, Bradley et al. [82] developed a methodology for the determination of 15 PAE diesters [DMP, DEP, diisopropyl phthalate, DAP, DBP, DIBP, DPP, di-*n*-hexyl phthalate (DHP), BBP, DCHP, DEHP, DnOP, DDP, DiDP and DiNP] in foodstuffs purchased in England (high-fat, high-carbohydrate and high-protein matrices). Of the 261 foodstuffs tested, PAE were confirmed to be present in 77 samples. The highest incidence was for DEHP (66 samples) with the highest concentration in an oil sample (6,447 µg/kg). However, the highest levels found were for the isomeric mixture DiNP in certain samples. In this study, additional PAE DIBP (bread, oils and fats, nuts and cereal), DBP (bread, oils and fats, nuts, meat products, cereal, fish and carcass meat), BBP (bread) and DEP (cereal) were reported to be present.

Cao et al. [83] investigated the occurrence of 20 PAE (DDP, di-*n*-undecyl phthalate, DMP, DEP, DEHP, DIBP, DBP, DPP, DHXP, di-(4-methyl-2-pentyl) phthalate, DCHP, BBP, DnOP, DNP, bis(2-*n*-butoxyethyl)phthalate, bis(2-ethoxyethyl) phthalate, bis(2-methoxyethyl) phthalate, dibenzyl phthalate, DPRP and DAP) in 159 composite food samples used in the 2013 Canadian Total Diet Study. Only six of the 20 PAE were detected, and the highest incidence was for DEHP (111 samples), followed by DBP (44 samples), (BBP) (32 samples), DIBP (27 samples), DEP (three samples) and DCHP (one sample). Levels of DEHP ranged from 14.4 to 714 ng/g but the highest levels were found in the vegetable and fruit samples. The presence of DEHP in some vegetable samples could be from packaging materials; however, contamination could be also from the environment. In general, the levels of DEP, DIBP, DBP, BBzP and DCHP were low. Table 4.4 shows conventional LC and GC methods reported for the analysis of PAE in foods.

Other studies of PAE determination in food simulants have also been reported. Moreira et al. [98] analysed the migration of DBP and BBP from plastic food containers into liquid food simulants (ultrapure water) during microwave operation; for this purpose, SPME was used prior to GC–MS analysis. Results showed that there was a greater migration of PAE in containers with a prolonged time of use, which is related to the increased heating time.

Perez-Outeiral et al. [99] developed a method based on ultrasound-assisted DLLME followed by solidification of floating organic drop and GC–FID for the determination of five PAE (DBP, BBP, DCHP, DEHP and DnOP) in food simulants [3% (w/v) acetic acid, 20% (v/v) ethanol] and liquid samples (water, vinegars, wines, soft drinks and sangria). The influence of different extraction parameters was investigated, including selection of extraction solvent, centrifugation and cooling conditions, dispersant, volume of extraction solvent, effect of pH, time and temperature. Results showed that time and temperature were not significant in the studied analytes; however, pH was found to be an important factor that might affect extraction efficiency as it affects the solubility of the analytes in the aqueous phase; as a result, the pH range 2–12 was selected.

Packaging made of recycled fibre materials, such as paper and paperboard, was found to be another source of PAE contamination in food. Gartner et al. [100] evaluated the migration of PAE in infant food packed in recycled paperboard, and the main contaminant in infant foods was DIBP (20.3 ng/g). However, PAE DIBP, DBP, DEHP and DiNP were also detected in food samples. The authors concluded that the application of appropriate secondary-packaging foil could prevent the contamination of foodstuff.

Table 4.4: GC and LC methods for PAEs determination in food products.

GC ethods							
Analytes	**Matrix**	**Detector**	**Sample preparation or extraction**	**Column/injection mode/oven temperature**	**LOD/LOQ**	**Results**	**Reference**
DMP, DEP, DBP, BBP, DEHP, DnOP	16 samples: olive oil	MS/MS	Gel permeation chromatography as clean-up step with cyclohexane: dichloromethane 7:3 as mobile phase	Column: 5 ms (30 m × 0.25 mm × 0.25 µm) Injection mode: Splitless Oven temperature: 100 °C (3 min), then ramped at 16 °C/min to 250 °C (28 min).	0.1–148 µg/kg/(0.2–182 µg/kg)	DBP, BBP and DEHP are the only PAEs present in the samples. DEHP was present in extra virgin olive oil (0.85 mg/kg) and in olive oil (1.45 mg/kg).	[67]
DMP, DEP, DBP, BBP, DEHP and bis(1-octyl) phthalate	Bottled mineral water samples	MS	Stir bar sorptive extraction with liquid desorption	Column: TRB-5MS (30 m × 0.25 mm × 0.25 µm) Injection mode: solvent vent Oven temperature: from 70 °C (2 min) at 25 °C/min to 150 °C, then at 3 °C/min to 170 °C, at 30 °C/min to 185 °C, at 3 °C/min to 195 °C, at 60 °C/min to 255 °C and then at 8 °C/min to 280 °C (4 min).	LOD: 0.15–0.60 µg/L	All PAEs were detected. DBP was the most abundant phthalate (0.35 µg/L) while DMP was under the LOD.	[84]

DBP	41 food samples (primarily dried foods such as cookies, pasta, oriental noodles, spaghettis) FCM: 69 paper packaging samples	*IT-MS/MS	Liquid-phase extraction coupled with ultrasonication	Column: Varian (30 m × 0.25 mm × 0.25 μm) Injection mode: Splitless Oven temperature: 80 °C (3 min) and increased to 280 °C at 25 °C/min (2 min).	0.003 mg /kg/(0.01 mg/kg)	DBP was detected in 4 imported (< LOQ 0.01–0.81 mg/kg) and two domestic (0.13 mg/kg and < LOQ < 0.01 mg/kg) food samples.	[85]
DMP, DEP, DBP, DEHP, BBP, DnOP	Wine (white, rose and red)	MS	Headspace solid-phase microextraction (HS-SPME)	Column: Varian CP8843 WCOT (30 m × 0.32 mm × 0.25 μm) Injection mode: Splitless Oven temperature: 60 C (1 min), then increased at a rate of 10 °C/min to 150 °C and (1 min), increased at a rate of 5 °C/min to 180 °C (2 min), then increased at a rate of 20 °C/min to 240 °C (6 min)	LOD: 0.0247–0.4004 μg/L	DBP was found, followed by DEHP (highest concentrations) and DEP. Total phthalate concentrations were between 2.7 and 15 μg/L	[52]

(continued)

Table 4.4 (continued)

GC ethods							
Analytes	**Matrix**	**Detector**	**Sample preparation or extraction**	**Column/injection mode/oven temperature**	**LOD/LOQ**	**Results**	**Reference**
DMP, DEP, DBP, DIBP, DPP, DEHP, DiNP, DIDP, DPrP, DCHP, BBP, DnOP	Oily foods 50 samples of oily foods packed in glass jars: The samples included oily products such as tomatoes in oil, olives in oil, vegetables in oil, tomato sauce and pesto.	MS/MS	Solid-phase extraction (SPE)	Column: ZB-5 ms (30 m × 0.25 mm × 0.25 μm) Injection mode: Splitless Oven temperature: 100 °C increased at 10 °C/min to 270 °C (4 min) increased at 25 °C/min to 285 °C (8 min).	LOQ: 0.05–2 mg/kg	DEHP was detected in 20 samples (0.1–6 mg/ kg). In 5 cases, the levels of DEHP were higher than the SML. DIBP was found in 4 samples (0.1–0.4 mg/kg). Other PAEs were not detected in any sample	[86]

DMP, DEP, DIBP, DBP, bis-2-methoxyethyl phthalate (MEP), bis-1,2-(4-methyl -2-pentyl) phthalate (1,2-MPP) Bis-1,3 (4-methyl -2-pentyl) phthalate (1,3-MPP), bis-2-ethoxy ethyl phthalate (EEP), di-*n*-amyl phthalate (DAmP) DHP, BBP, bis-2-butoxyethyl phthalate (BBEP), DCHP, DEHP, DnOP, DNP	Olive oil	IT/MS	HS-SPME	Column:ZB-5 MS (30 m × 0.25 mm × 0.25 μm) Oven temperature: 70 °C (2 min), raised to 200 °C (6 °C/min, 5 min), raised to 295 °C (4 °C/min, 8 min)	LOD: 0.02–0.05 mg/kg	Samples kept in PET containers shown DIBP (119–216 μg/kg), DBP (97–175 μg/kg), BBP (173–211 μg/kg) and DEHP 543–840 μg/kg), samples kept in glass containers: BBP (87–112 μg/kg) and DEHP (198–237 μg/kg) were detected.	[87]
DMP, DEP, DBP, DAmP, DnOP	Soybean milk	MS	Molecularly imprinted solid-phase extraction (MISPE)	Column: MS DB-5 MS (30 m × 0.25 mm, 0.25 μm) Injection mode: splitless Oven: 150 °C (1 min), 150–170 °C at 30 °C/min, then to 300 °C (40 °C/min, 3 min)	0.013–0.022 μg/mL/ (0.040–0.056 μg/mL)	DEP, DBP and DnOP were found with the concentrations of 7.6, 8.8 and 9.6 μg/L, respectively.	[58]

(continued)

Table 4.4 (continued)

Analytes	Matrix	Detector	Sample preparation or extraction	Column/injection mode/oven temperature	LOD/LOQ	Results	Reference
GC ethods							
DMP, DEP, DBP, DIBP, BBP, DEHP, di-*n*-hexyl phthalate (DNHP), DCHP, DnOP	78 food samples: beverages, condiments, milk and milk products, cooking oil, seafood, meat and meat products and others.	MS	Liquid samples (except for milk): LLE with hexane. Solid samples (including milk): LLE with hexane: acetone (1:1 v/v). For cooking oils: LLE with hexane-saturated acetonitrile	Column: DB-5 (30 m × 0.25 mm; 0.25 μm). Injection mode: not available Oven temperature: 80 °C (1.0 min), raised to 180 °C at 12 °C/min (1.0 min), increased to 230 °C at 6 °C/min, then increased to 270 °C at 8 °C/min (2.0 min) and finally ramped up to 300 °C at 30 °C/min (12 min).	LOQ: 2–10 ng/g	DMP, DEP, DBP, DIBP, BBP and DEHP were frequently detected (> 60%), whereas DNHP, DCHP, and DnOP were found in fewer than 16% of the samples. DBP and DIBP were found in >94% of the food samples (0.011–572 ng/g). while DEHP (<LOQ, i.e. <0.2 to 762 ng/g) and DEP (<LOQ, i.e. <0.04 to 22.4 ng/g) was found in 82% and 81% of the samples, respectively.	[88]

DMP, DEP, BBP, DBP, DIBP, DEHP, DNHP, DCHP, DnOP	Beverages, milk, other dairy, fish, fruits/ vegetable, grain, beef, pork, poultry, meat and meat products, vegetable oils, condiments, infant foods.	MS	Liquid samples (without lipids): extraction with hexane. Solid foods: extracted with acetone:hexane (1:1) Clean up by column chromatography. Purification of extracts: glass column packed with 7 g Florisil 60–100 mesh.	DB-5 (30 m × 0.25 mm 0.25 µm) Injection mode: Splitless Oven temperature: 80 °C (1 min), raised to 180 °C at 12 °C/min (1 min), increased to 230 °C at 6 °C/min then to 270 °C at 8 °C/min (2 min) and to 300 °C at 30 °C/ min (12 min).	LOD: 0.2–3.7 ng/kg	DEHP was detected in 74% of food samples, following DEP (57%), DiBP (55%), BBP (54%), DMP (37%), DBP (31%) and DCHP (6%) DEHP had the highest concentration (4 ng/mL for beverages to 300 ng/g for pork)	[89]
DEP, DMP, DIBP, DBP, BBP, DEHP	Carbonated and noncarbonated mineral waters bottled in PET containers	a. MS/MS b.**Py-GC /MS	Liquid–liquid extraction (LLE)	Column:a.SGE (30 m × 0.25 mm × 0.25 µm). b. DB-1701 (30 m × 0.25 mm × 0.25 µm) Injection mode: not available Oven temperature: (a) 100 °C (1 min), then heated up to 300 °C (20 °C/min, 5.5 min) (b) 50 °C (1 min) to 280 °C at a rate of 30 °C min	LOQ: 3.0–22.2 ng/ L	Noncarbonated samples: DIBP (<3.0 ng/L to 0.2 µg/ L) Di-*n*-butyl-phthalate (<6.6 ng /L to 0.8 µg/ L). Benzyl-butyl phthalate (<6.0 ng /L to 0.1 µg/ L) and DEHP(<16.0 ng/L to 1.7 µg/ L). Carbonated mineral water samples: none of the phthalate could be detected. DEHP was the most abundant phthalate in the investigated samples.	[90]

(continued)

Table 4.4 (continued)

GC ethods							
Analytes	**Matrix**	**Detector**	**Sample preparation or extraction**	**Column/injection mode/oven temperature**	**LOD/LOQ**	**Results**	**Reference**
DMP, DEP, DBP, BBP, isobutyl phthalate (IBP), DEHP	Hydroalcoholic food beverages	FID and IT/MS	SPE with Amberlite XAD-2 adsorbent	Column: SE-54 (30 m × 0.25 mm × 0.24 μm) Injection mode: Splitless Oven temperature: 100–300 °C at 10 °C/min	1.21–2.51 pg/μL/ (2.42–5.03 pg/μL)	Wine: DBP and DEHP (levels above LOQs), other PAEs are below the relative LOQs. Vodka liqueur: DBP, BBP, IBP and DEHP show levels above the relative LOQs. DEHP is the only PAE always present in all the samples investigated, except in a red wine sample	[36]
DBP, DEHP	Chicken soup	FID	Magnetic microsolid-phase extraction	Column: DB-5 (30 m × 0.25 mm × 0.25 μm) Oven temperature: 90 °C (2 min) before it increased to 300 °C (5 min) at 20 °C/min.	26.3–36.4 ng/ mL/ (88–121 ng/mL)	Only DEHP was found in clear chicken soup samples in the range 0.02–0.07 μg/mL	[38]

LC methods							
Analytes	**Matrix**	**Detector**	**Sample preparation or extraction**	**Column/mobile phase**	**LOD/LOQ**	**Results**	**Reference**
DBP, BBP, DEHP, DiNP, DIDP	33 milk and milk products including infant formulas (raw milk, pasteurised and homogenised milk, yoghurt with fruit, reconstituted infant formula from different part of world, liquid infant formula)	***ESI-MS /MS	Liquid samples: mixture of tert-butyl methyl ether and hexane. Fats: liquid–liquid extraction into acetonitrile	Column: Luna C5(20 mm × 50 mm × 5 μm) Isocratic conditions Mobile phase: 2.0% v/v water in methanol/acetonitrile (1 + 1). Post-column reagent contained 1.0 mM ammonium acetate in 80% methanol.	LOD: 4–9 μg/kg	DEHP concentrations were clearly dependent on the fat content (levels generally low, except in a few infant formulas – more than 100 mg/kg)	[91]
DMP, DEP, DPrP, BBP, DCHP	Fruit jellies	MS	QuEChERS Dispersive solid-phase extraction	Column: C8–3 (2.1 × 150 mm × 5 μm) Gradient mode Mobile phase: methanol–water mixture 0–4 min (50–80% methanol); 4–10 min (80–90% methanol); and then return to 50% methanol in 0.1 min; finally, 50% methanol (5 min)	0.09–3.68/(0.28–11.15 ng/mL)	BBP was found in all the samples (2.9–14.7 mg/kg). DEP was found in 3 samples (0.49–1.2 mg/kg) DMP, DPrP and DCHP were not found in all the samples.	[92]

(continued)

Table 4.4 (continued)

LC methods							
Analytes	**Matrix**	**Detector**	**Sample preparation or extraction**	**Column/mobile phase**	**LOD/LOQ**	**Results**	**Reference**
DMP, DEP, BBP, DBP, DEHP, DnOP, DiNP, DIDP	Cooking oil and mineral water	UV	SPE by nylon 6 nanofibre mat as sorbent	Column: C18 (150 mm × 4.6 mm × 5 μm) Gradient mode Mobile phase: acetonitrile in water (60–100% acetonitrile for 0–15 min followed by 100% acetonitrile for 15–35 min) for 40 min.	LOD: Oil:0.020–0.150 ng/mL Wáter: 0.001–0.020 ng/mL	DBP (shown relatively high abundance in both matrices). BBP, DEHP and DnOP (shown considerable levels in the blank controls)	[93]
DMP, DEP, DBP, DEHP, DnOP, BBP	Orange juice packaged in polyvinyl chloride (PVC) bottle.	UV	SPE	Column: C18 (250 mm × 4.6 mm × 5 μm) Gradient mode Mobile phase: mixture of acetonitrile and water 75–85% of acetonitrile in 5 min, then linear gradient from 85% to 100% of acetonitrile in 10 min and finally isocratic conditions with 100% acetonitrile for 6 min, returned to initial composition at 23 min, and held for 5 min.	LOD: 2.6–13.8 ng/mL	DEP and DEHP in orange juice samples would increase with the storage time and reach up to 0.385 μg/mL and 0.662 μg/mL.	[94]

DMP, DEP, DBP, DPP, DCHP, DnOP, DIDP, diisooctyl phthalate (DIOP), BBP	Milk products	ESI-MS-MS	LLE	Column: Hypersil Gold Column (50 mm × 2.1 × 1.9 μm) Gradient mode Mobile phase: A: methanol. B: 5 mmol ammonium acetate solution + 0.1% v/v formic acid (0–2 min) 20% A, 80% B (2–3.5 min) 20–95% A, 80–5%B; (3.5–11 min) 95% A, 5%B; (11–12 min) 95–20% A, 5–80% B; (12–13.5 min) 20% A, 80% B.	5–25/(17–83 μg/kg)	DBP was found to be more than 100 μg/kg in all this milk products. The commercial milk samples showed common existence of DMP. DIDP was not detected.	[35]
DEHP	Milk powder and soft drinks	ESI-MS	SPE	Column: C8 (4.6 mm × 150 mm × 5 μm) Gradient mode Mobile phase A: 1 L water + 0.5 mL 100% glacial acetic acid + 10 mg sodium acetate trihydrate; B: 900 mL methanol + 100 mL of water + 100% glacial acetic acid + 10 mg sodium acetate trihydrate. 60% A + 40% B, 0.5 mL/min (9 min); at 9.1 min 0% A + 100% B a flow rate of 0.7 mL/min (14 min).	Drinks: 13/(40 ng/L) Milk powder: 600/(8800 ng/kg)	DEHP was present in soft drinks in a range of 0.03–9.05 ng/L. In milk powder in a range of 5,300–25,100 ng/kg	[95]

(continued)

Table 4.4 (continued)

LC methods							
Analytes	**Matrix**	**Detector**	**Sample preparation or extraction**	**Column/mobile phase**	**LOD/LOQ**	**Results**	**Reference**
DMP, DEP, DIBP, DNBP, BBP, DEHP, DnOP, DiNP, DIDP	Wine	MS/MS	Filter (0.2 µm) No sample extraction	Column: C18 (75 mm × 2 mm × 4 µm) Gradient mode Mobile phase A: water containing 10 mM ammonium acetate B: methanol 300 µL/mL/min, a linear gradient of solvent B from 50% to 95% in 3 min, held 95% (9 min), to 100% (0.5 min) and held at 100% (7.5 min)	White wine: 0. 5–4.8/(1.6–14.6 µg/L) Red wine: 0. 6–8.8/(1.7–26.6 µg/L)	Concentrations of DMP, DEP, DnOP, DEHP, DiNP and DIDP in all samples – lower than either their LODs or LOQs. DIBP and DnBP were most commonly found, followed by DEHP or BBP. BBP was found in 4 red wines up to 6.3 µg/L. DnBP was detected in all the wines at concentrations up to 9.3 µg/ L .DiBP – detected in all red wines up to 10.7 µg/L.	[96]
Diphenyl phthalate (DPhP), DMP, DEP, DBP, BBP, di-*n*-propyl phthalate (DnPP).	Palm oil	DAD	Online solid phase extraction (SPE)	Column:C18 (150 mm × 4.6 × 5 µm) Isocratic and gradient elution Mobile phase: mixture of ultrapure water and 10 mM methane sulphonic acid (MSA), acetonitrile (ACN) and ethyl acetate	3 µg/L/(10 µg /L)	Only DBP and DMP were identified in samples with concentrations less than 1 mg/L.	[97]

DPhP, bis(2-*n*-butoxyethyl) phthalate (DBEP), bis(2-ethoxyethyl) phthalate (DEEP), DEHP, bis(2-methoxyethyl) phthalate (DMEP), bis(4-methyl-2-pentyl) phthalate (BMPP), BBP, DBP, dicyclohexyl phthalate (DCHP), hexyl 2-ethylhexyl phthalate (HEP), DPP, DEP, di-*n*-hexyl phthalate (DHXP), DIBP, DMP, DNP, DnOP, DPrP, DAP, DIDP	623 samples: milk-based products, distilled liquor, wine, beverage, grain, meat, oil, biscuit (cookie), and canned food	MS/MS	Liquid samples were exacted by acetonitrile, while solid samples were prepared by QuEChERS or glass-based SPE methods	Column: Poroshell 120 EC-C18 (100 × 4.6 mm × 2.7) Gradient mode Mobile phase: methanol – 0.1% formic acid water mixture 0–12.0 min, 40–100% methanol; 12.0–15.0 min, 100% methanol; then return to 40% methanol in 0.1 min; finally, 15.1–18.0 min, 40% methanol.	0.8–15/(10–100 µg/kg)	DEHP was found in almost all samples (0.02–2,685 mg/kg) In distilled liquor samples, DBP and DEHP were detected quite frequently as (0.3 and 1.5 mg/kg, respectively) Slightly higher levels of DBP, DEHP and DiNP were found in soft drinks.	[37]

*Ion trap; **Pyrolysis; ***Electrospray ionisation
FCM, food contact materials.

On the other hand, in 2014, Gao et al. [101] developed an SPE–HPLC method for the separation and quantification of DMP, DEP, DPRP, BBP, DIBP, DCHP, DAP, DHP, DnOP and DEHP released from food paper packaging materials using food simulants [distilled water, 3% acetic acid (w/v), 10% ethanol (v/v) and 95% ethanol (v/v)].

The main problem associated with PAE analysis was an issue with the blank. The analysis of blanks is fundamental to avoid false-positives or overestimated results. Due to its presence as an environmental contaminant, PAE contamination may be encountered during the analytical procedure.

In order to minimise this problem, several precautions are taken during sample preparation and chromatographic analysis. The cleaning procedure of laboratory equipment is strictly controlled prior to use, all glassware used in the procedure must be rinsed and thermal treatment at 400 °C applied. In the same way, the use of plastic material is avoided and it is recommended that a sample preparation procedure is comprised of minimal steps in order to minimise potential sources of PAE contamination [76, 82, 102].

4.2 Alternative plasticisers

Other than PAE, a number of substances have been identified as alternative plasticisers, which offer different properties to polymeric materials. These alternatives include citrates, sebacates, adipates, phosphates and epoxy esters.

Adipates can be blended with PAE for PVC applications, and the desirable properties of adipates include low viscosity, higher gelation and fusion temperature. One of the most common adipates is DEHA, which is an organic compound prepared via the esterification of adipic acid with 2-ethyl hexanol [103]. DEHA is widely used as a plasticiser in flexible PVC films as well as in coatings, adhesives and printing inks. DEHA can be used as a substitute for PAE in PVC applications, including films employed in food wrapping material. In this sense, people may be exposed to DEHA through its migration from FCM. A TDI of 0.3 mg/kg has been established for DEHA [103–105].

Citric acid esters are mainly used as plasticisers in polymers such as PVC and cellulose acetate. Citric acid is the starting material for a number of citrate plasticisers. One of the most common citrates is acetyl tributyl citrate (ATBC), a compound compatible with PVC and other food contact polymers, and it can be used as a replacement for DEHA or DEHP. Other citrates include acetyl triethyl citrate and tributyl citrate [103, 106].

Sebacates are alternative plasticisers having good compatibility with a PVC matrix. Dioctyl sebacate is one of the most important sebacates and is widely used as a plasticiser because it provides excellent low-temperature flexibility to elastomers. Dibutyl sebacate (DBS) is other sebacate, principally used as a plasticiser in film coatings (10–30%) [103].

Phosphate esters are used as plasticisers in order to enhance the flame retardancy of flexible PVC. Examples of this compound include phosphoric acid, triethyl ester (TEP) and phosphoric acid and triphenyl ester (TPP) [103].

Epoxidised soybean oil (ESBO) is another alternative plasticiser, resulting from the epoxidation reaction of soybean oil. This compound is widely used as a plasticiser and heat stabiliser in PVC formulations including FCM. Gaskets of the lids for glass jars may also contain ESBO in their formulation [107].

Terephthalate esters have emerged as a safe alternative for traditional plasticisers. In particular, DEHT is used as a substitute for DEHP in several applications, including medical and FCM [108].

Another alternative plasticiser with similar technical properties to DEHP is diisononyl cyclohexane-1,2 dicarboxylate (DINCH), a chemical compound with a specific migration rate lower than DEHP and is used in a range of applications (medical products, toys, food packaging materials, printing inks and protective coatings, among others) [109]. Table 4.5 includes some alternative plasticisers and their applications.

Just like PAE, alternative plasticisers used in food contact applications are also involved in RASFF notifications, and since 2010, 46 notifications have been reported. ESBO was involved in 36 notifications (see Table 4.6). Most ESBO notifications are in relation to the migration from lids of jars containing different food products. Other plasticisers involved in the notifications were DINCH, DEHA, ATBC and DOTP [77].

The migration of plasticisers from FCM has been determined mainly by GC–FID or MS [78, 115]. Other methods reported in the literature for the determination of migrating plasticisers include LC, pyrolysis GC–MS, ultraperformance LC (UPLC)–MS/MS and online LC–GC with FID [83–85, 116–118].

Goulas et al. [119] studied the migration of DEHA from food-grade PVC film into hard and soft cheeses. DEHA migration was determined by GC–FID. High levels of DEHA were found in all three cheese samples tested (345.4 mg/kg for Kefalotyri cheese, 222.5 mg/kg for Edam cheese and 133.9 mg/kg for Feta cheese).

In 2007, a kinetic study of the migration of DEHA and ATBC from PVC film into sweetened sesame paste (halva) samples was carried out. A PVC film containing DEHA (5.3%) and ATBC (3.0%) was used to wrap halva samples, and the determination of both plasticisers was achieved using GC–FID. The equilibrium content of DEHA in halva (81.4 mg/kg halva) corresponded to a loss of 54.7% DEHA from the PVC film, while the equilibrium content of ATBC in halva was 36.1 mg/kg, corresponding to a loss of 42.7% ATBC from the PVC film [120].

Fankhauser-Noti et al. [107] investigated the migration of ESBO and other plasticisers in food packed in glass jars with metal closures. ESBO and epoxidised linseed oil (ELO) were analysed by LC–GC–FID while the other PVC additives/plasticisers were determined using GC–MS; 64% of the gaskets contained ESBO as the principal plasticiser, followed by PAE (22%). Concentrations found in food products were 1,170 mg/kg for ESBO, 270 mg/kg for DiNP, 740 mg/kg for DiDP, 825 mg/kg

Table 4.5: Alternative plasticisers and their applications.

Structure	Name	Acronym	CAS no.	Formula	Molecular weight	Application	SML[a] (mg/kg)	Reference
Adipates: Alcohols in the C6–C10 range can be esterified with adipic or azelaic acid to produce a range of adipate plasticisers								
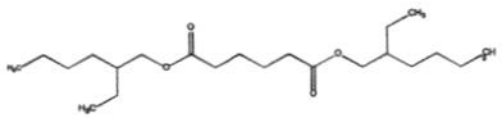	Bis(2-ethylhexyl) adipate	DEHA	103-23-1	$C_{22}H_{42}O_4$	370.57	Widely used in flexible PVC for food packaging applications. Combines well with the majority of polar thermoplastics. Other applications: cosmetics, pharmaceuticals, medicals, lacquers, bottle caps, paper and aluminium foil coatings, hair spray.	60 *	[1,16]
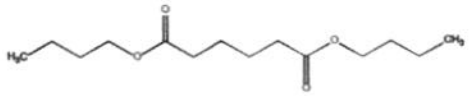	Dibutyl adipate	DBA	105-99-7	$C_{14}H_{26}O_4$	258.35	Plasticiser for PVC, ethyl and nitrocellulose, polystyrene.	–	[16], 18
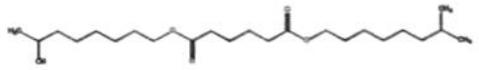	Diisononyl adipate	DINA	33703-08-1	$C_{24}H_{46}O_4$	398.62	Plasticiser for PVC and its copolymers and rubbers.	–	[110]

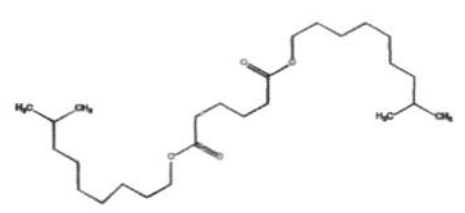	Diisodecyl adipate	DIDA	27178-16-1	$C_{26}H_{50}O_4$	426.67	Used primarily in flexible PVC applications. Recommended for cold resistant articles: cable and wire, shoes, films, sheet, synthetic leather.	–	[110]
Citrates: Prepared by esterification of citric acid and 3 mol of alcohol (*n*-butanol or *n*-hexanol)								
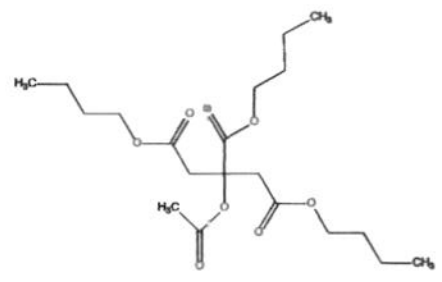	Acetyl tributyl citrate	ATBC	77-90-7	$C_{20}H_{34}O_8$	402.48	Main plasticiser of PVC Plasticiser for films, in particular vinyl chloride (VC)-vinylidene chloride (VDC) copolymer and cellulose. Others: cosmetic products, toys, vinyl, adhesives, medical devices, pharmaceutical tablet coatings.	60*	[16, 111]
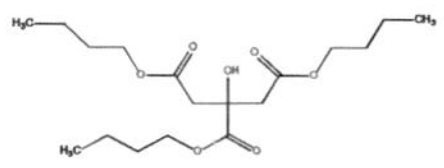	Tributyl citrate	TBC	77-94-1	$C_{18}H_{32}O_7$	360.44	Used as a plasticiser in PVC, cellulose acetate, cellulose nitrate. Solvent for nitrocellulose, lacquers in contact with food.	–	[16,-112]

(continued)

Table 4.5 (continued)

Structure	Name	Acronym	CAS no.	Formula	Molecular weight	Application	SML[a] (mg/kg)	Reference
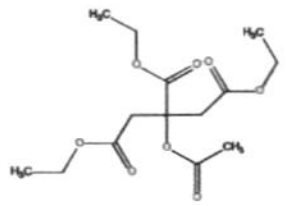	Acetyl triethyl citrate	ATEC	77-89-4	$C_{14}H_{22}O_8$	318.31	Mainly used as a plasticiser in food packaging materials.	–	[112]
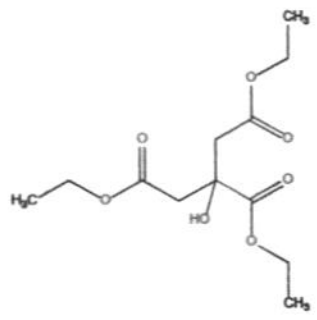	Triethyl citrate	TEC	77-93-0	$C_{12}H_{20}O_7$	276.29	Plasticiser in the production of polyvinyl acetate, cellulose citrate and cellulose acetate (food wraps)	60 *	[112]
Sebacates: esters of 1,8-octanedicarboxylic acid – a ten-carbon aliphatic acid								
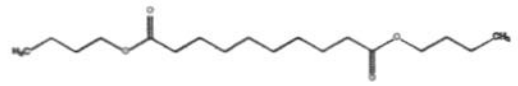	Dibutyl sebacate	DBS	109-43-3	$C_{18}H_{34}O_4$	314.46	Plasticiser in the production of PVC and its copolymers, acetobutyrate cellulose, polyvinyl butyral, chlorinated rubber, polyvinylidene chloride. Used especially for polyisopropene, food contact, medical and pharmaceutical plastics.	60 *	[16, 113]

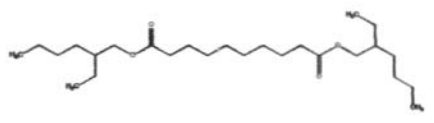	Dioctyl sebacate	DOS	122-62-3	$C_{26}H_{50}O_4$	426.67	Low-temperature-resistant plasticiser to produce PVC and PVA elastomers Compatible with nitro- and ethyl cellulose and with PVC.	–	[16]
Phosphate esters: prepared from $POCl_3$ and monofunctional alcohols or alkylated phenols.								
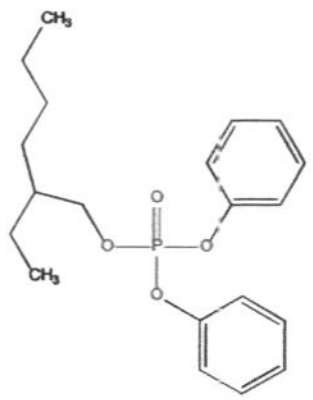	Phosphoric acid, diphenyl 2-ethylhexyl ester	DPEHP	1241-94-7	$C_{20}H_{27}O_3P$	362.41	Used in PVC and vinyl polymers Mixes well with cellulose acetobutyrate, polymethyl methacrylate, polyvinyl butyral, among others.	2,4	[112]
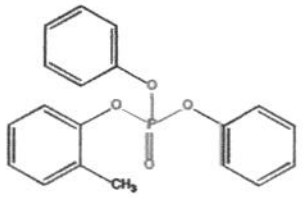	Phosphoric acid, diphenyl tolyl ester		26444-49-5	$C_{19}H_{17}O_4P$	240.33	Plasticiser and flame retardant in the production of PVC, polystyrene, polycarbonate, butadiene, cellulose acetate and acetobutyrate and other food contact polymer materials.	–	[112]
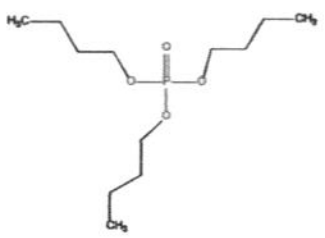	Phosphoric acid, tributyl ester	TBP	126-73-8	$C_{12}H_{27}O_4P$	266.32	Used as a plasticiser in nitrocellulose, plastics and vinyl resins intended for contact with food.	–	[112]

(continued)

Table 4.5 (continued)

Structure	Name	Acronym	CAS no.	Formula	Molecular weight	Application	SML[a] (mg/kg)	Reference
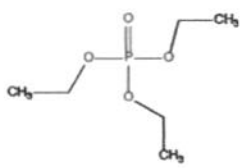	Phosphoric acid, triethyl ester	TEP	78-40-0	$C_6H_{15}O_4P$	182.16	Plasticiser in the production of food contact plastics.	–	[112]
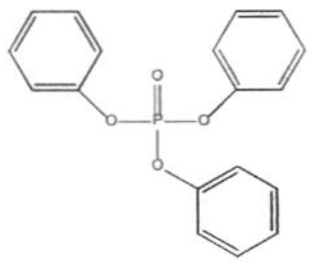	Phosphoric acid, triphenyl ester	TPP	115-86-6	$C_{18}H_{15}O_4P$	326.31	Plasticiser in the production of cellulose acetate articles and in lacquers manufacture.	–	[112]
Epoxy esters: manufactured by reaction of performic or peracetic acid on soybean or linseed oil.								
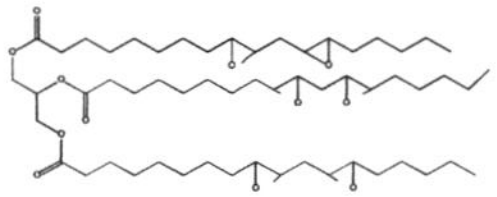 Epoxidised linolein, a major component of ESBO	Epoxidised soybean oil	ESBO	8013-07-8	NA	NA	Plasticising properties for PVC and chlorinated rubber. PVC plasticiser Gaskets of the lids for glass jars.	60* 30**	[16]

Terephthalates esters: similar to PAEs (two adjacent ring substitutions occupying para-positions instead of ortho-positions)

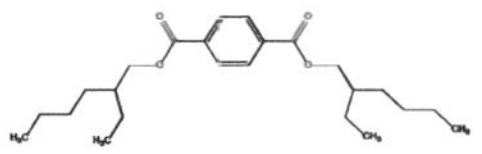	Di(2-ethylhexyl) terephthalate	DEHT	6422-86-2	$C_{24}H_{38}O_4$	390.56	Used as a plasticiser in flexible PVC. Others: medical devices, toys, childcare articles and food contact applications.	60 *	[111, 114]

* SML(T) (group restriction).
** In the case of PVC gaskets used to seal glass jars containing infant formulae.
[a] Commission regulation (EU) no. 10/2011.
N/A, not available.

Table 4.6: RASFF notifications of plasticisers since 2010.

	RASFF notifications						
	Year						
Compound	**2010**	**2011**	**2012**	**2013**	**2014**	**2015**	**2016**
ESBO	4	10	11	1	1	2	3
ESBO + (cyclohexanedicarboxylic acid diisononyl ester – DINCH	2		1				
DINCH	1						
ATBC + DEHA + DINCH			1				
(ESBO) + 2-ethyl hexyl palmitate and stearate (EHOL)			1				
Bis(2-ethylhexyl) terephthalate (DOTP)					5	2	1

for DEHP and 180 mg/kg for DEHA. Furthermore, elevated concentrations of DINCH, 2-ethylhexyl palmitate, stearate and ELO were found.

Ezerskis et al. [121] determined polyadipates (PAD) and other plasticisers in PVC gasket seals and in fatty food (pesto, tomato sauces, olive oil and olives in oil) by GC–MS. ESBO was the principal plasticiser in gaskets (15–42%), while DiDP was found in three samples as the main plasticiser (37–41%). DBS, DEHP, DEHA and di-(2-ethylhexyl) sebacate (DEHS) were also found in the gasket samples. In the case of food samples, ESBO was detected with an average concentration of 61.3 mg/kg. DEHP was detected in six food samples ranging from 2.5 to 8.7 mg/kg. DEHA, DEHS and DiDP were nondetectable in all the food samples analysed.

Suman et al. [122] developed a methodology for the determination of ESBO–chlorohydrins (as methyl esters) in food sauces using UPLC–ESI–MS/MS. There were positive results for 18-*E*-OHCl chlorohydrins, whereas no traces of 18-2-OHCl chlorohydrin were found. Table 4.7 shows LC and GC methods reported for the analysis of plasticisers in foods.

The migration of plasticisers into food simulants has also been reported in several studies [79]. The effect of gamma radiation on the migration of DEHA and ATBC plasticisers from PVC film into the food simulant isooctane was studied by Zygoura et al. [133]. Results showed that the migration level increased with increasing irradiation dose and contact time; in the same way, DEHA migrated rapidly into isooctane compared with ATBC.

Bueno-Ferrer et al. [134] evaluated the migration of ESBO in food simulants (fat and aqueous); for this purpose, samples of PVC were prepared with three different concentrations of ESBO (30%, 40% and 50%) . Results showed that the migration

Table 4.7: LC and GC methods for phthalates/other plasticisers determination in food and food contact materials.

Analytes	Type of sample	Analytical technique	Sample preparation and/or extraction procedure	Column/mobile phase or oven T°	LOD/LOQ	Results	Reference
DBP, BBP, DEHP, DiNP, di-*n*-alkyl adipate (DAA) DEHA, diisononyl adipate (DINA), dibutyl sebacate (DBS), ATBC, diacetyl lauroyl glycerol (DALG)	Food: 93 samples (beverages, fat and oil, dairy products, refreshments, fast food, instant food, baby food).	GC-MS	Retort-pouched foods, pizza and gyudon: extraction with ACN Sake and wine: LLE with hexane Butter, margarine, fat spread: extraction with acetone Florisil and Bondesil PSA dual-layer column was used for the clean-up of extracts	Column: DB-5MS (0.25 mm × 30 m × 0.25 μm) Injection mode: Splitless Oven temperature: 50 °C (1 min), then at 10 °C/min to 270 °C (27 min).	LOD: 0.0002–0.28 μg/g LOQ: was set at twice the LOD	DEHP was found in baby food sample (source of contamination was presumed to be disposable gloves). DALG was detected in the other baby food samples (did not originate as contamination from plastics but was added as a food additive). ATBC was detected in beverages (bottled sake in levels of around 3–7 μg/g)	[123]

(continued)

Table 4.7 (continued)

Analytes	Type of sample	Analytical technique	Sample preparation and/or extraction procedure	Column/mobile phase or oven T°	LOD/LOQ	Results	Reference
DMP, DEP, DPrP, DBP, BBP, DCHP, DEHP, DIDP, diacetyl lauroyl glycerol (DALG), DiNP, DEHA, di-hexyl phthalate (DHP). hexyl octyl decyl adipate (HXODA), heptyl nonyl adipate (HNA) and diisononyl adipate (DINA), ATBC, DBS	FCM: 98 samples of cap-sealing resins for bottled foods.	GC-FID GC/MS	LLE with hexane	Column: HP-5 (30 m × 0.32 mm × 0.5 μm) Injection mode: Split (20:1) Oven temperature: 150 °C (1 min) to 250 °C at 20 °C/min (24 min).	LOD: – LOQ: –	Period 1997–1999: DEHP was found in domestic and imported samples, followed by DEHP. Period 1993–1999: diacetyl lauroyl glycerol (DALG) was detected in domestic samples, DIDP and DiNP were detected only in imported samples.	[124]
DEHP, DEHA	Food: curry paste.	GC-FID	Ultrasonic and solid-phase extraction	Column: HP-5MS (30 m × 0.25 mm 0.25 μm) Oven: 100 °C (1 min), then ramped up at 20 °C/min to 300 °C (2 min)	27–30/ (90–100 μg/L)	DEHP (0.12–0.61 μg/g) and DEHA (4.0–26.4 ng/g) were found in the samples.	[125]

DMP, DEP, DEHP, DIBP, DBP, DHP, BBP, DEHA, DnOP	Food: bottled water samples.	GC/MS	Solid-phase microextraction (SPME)	Column: DB-5MS (3.0 m × 0.25 mm, 0.25 μm) Oven temperature: 50 °C (5 min), raised to 280 C at 15 °C/min (5 min)	0.003 – 0.085 μg/L/–	DMP, DHP, BBP, DEHA and DnOP were not detected in any samples. Concentrations of the other PAEs were low (around sub-ppb) except for DBP (1.72 μg/L)	[126]
ESBO, DIDP, DiNP, DEHP, DEHA, DBS	FCM: PVC gaskets Food: 19 samples of food in glass jars with twist closures, oily products such as tomatoes, olives, garlic and shrimps in oil, herring in pickle and dressing, milk products, soft spreadable cheese, sauces, peanut butter, and infant food (mixed dishes of vegetables, rice and meat).	LC-GC-FID (ESBO) GC-MS (PAEs)	Homogenisation. ESBO: transesterification PAEs: injector-internal thermal desorption	–	LOD: 1–3 mg/kg (ESBO) below 1 mg/kg (PAEs)	ESBO was found in seven of the food samples at concentrations from 6 to 100 mg/kg DIDP and DiNP were found in seven samples at 6–173 mg/kg	[127]

(continued)

Table 4.7 (continued)

Analytes	Type of sample	Analytical technique	Sample preparation and/or extraction procedure	Column/mobile phase or oven T°	LOD/LOQ	Results	Reference
DMP,DEP, DBP, DIBP, DPP, DEHP, DiNP, DIDP, DPrP, DCHP, BBP, DnOP. Plasticisers: ATBC, DEHA, DBS, di-isononyl cyclohexane 1,2-dicarboxylate (DINCH)	Food: tomato sauce, pesto sauce, and sunflower seed oil.	GC/MS (ATBC, DEHA, and DBS) GC/MS/MS (PAEs and DINCH)	Solid-phase extraction (SPE)	Column: ZB-5 ms (30 m × 0.25 mm × 0.25 µm) Injection mode: Splitless Oven temperature: 100 °C, increased by 10 °C/min to 270 °C (4 min), increased by 25° C/min to 285 °C (8 min).	LOQ: 0.05–2 mg/kg	Recoveries were in the range of 71–105% with RSD ≤ 12%. LOQ values were 0.050–0.1 mg/kg, except for DiNP, DIDP, and DIDP (2 mg/kg).	[128]
DEHA	Food: 26 samples: cheese, meat, poultry, fish and fast foods	GC/MS	Dispersive SPE	Column: DB-5MS (30 m × 0.25 mm × 0.25 µm) Injection mode: – Oven temperature: 50 °C (2 min) and then 280 °C (6 min at 15 °C/min)	LOD: 0.02–0.2 mg/g	DEHA was detected in most of the meat, poultry and fish composite samples, with the highest concentration found in ground beef (11 µg/g)...	[129]

DEHA, DMP, DEP, DBP, DIBP, BBP, DHP, DEHP, DnOP	Food: 118 samples: meat (beef, pork, and chicken), fish and cheese packaged mostly in cling films	GC/MS	Dispersive SPE	Column: DB-5MS (30 m × 0.25 mm × 0.25 μm) Injection mode: – Oven temperature: 50 °C (2 min) then increasing to 280 ° C (6 min at 15 °C/min)	LOD: DMP, DHP, and DOP: 0.0016–0.0034 μg/g (DEP, DiBP, DBP, BBP, DEHA, DEHP 0.027–0.43 μg/g	DEHA – detected in samples packaged in DEHA plasticised cling films: cheese (average: 203 μg/g), beef (average 6.3 μg/g), pork (average 9.1 μg/g), chicken (average 2.5 μg/g), fish (average 5.9 μg/g). DEHP was detected in a few cheese samples (likely due to environmental contamination). The other 7 PAEs were not detected in any of the food samples.	[130]

(continued)

Table 4.7 (continued)

Analytes	Type of sample	Analytical technique	Sample preparation and/or extraction procedure	Column/mobile phase or oven T°	LOD/LOQ	Results	Reference
DEP,BBP, DEHA, DnOP, DCHP, DIBP, DBP, DEHP	Food: spices and roasted chicken meat stored in plastic bags.	GC/MS	SPME fibre cooled by liquid nitrogen	HP-5MS (30 m × 0.25 mm × 0.25 μm) Injection mode: Splitless Oven temperature:120 °C (3 min), increased to 190 °C at a rate of 20 °C/ min (4 min), increased to 250 °C at a rate of 2 °C/min, maintained for 1 min, further increased to 280 °C at a rate of 20 °C/min and finally maintained for 1 min.	LOD 0.01–0.18 μg/kg/LOQ 0.07–0.26 μg/kg	DIBP and DBP were found in most samples. However, the presence of PAEs in meat samples was due to the addition of spices during meat preparation.	[98]

DMP, DEP, DBP, BBP, DEHP, DnOP, DEHA	Food: 15 samples of beer	GC-MS	SPME	Column: HP-5-ms (30 m × 0.250 mm × 0.25 μm). Oven temperature: 70 °C (5 min) then increased to 250 °C at 20 °C/min (half a minute) then increased to 320 °C at a rate of 10 °C/min(5 min)	0.006–0.588 μg/L/(0.020–1.959 μg/L)	DEHP was the most commonly found (93% samples). PAEs were found in all samples at concentrations ranging from as low as 0.05 μg/L (for DEHP) to 37.14 μg/L (for DBP) and from 1.01 to 61.56 μg/L for total phthalates. DEHA was found in one sample (0.48 μg/L)	[131]

(continued)

Table 4.7 (continued)

Analytes	Type of sample	Analytical technique	Sample preparation and/or extraction procedure	Column/mobile phase or oven T°	LOD/LOQ	Results	Reference
Di-*n*-alkyl adipate (DAA), DINA, DALG	Food: rapeseed oil, Japanese radish, pumpkin, pineapple, melon, cooked radish with fried tofu, meat sauce, hamburger, potato salad, croquette, minced chicken white meat, minced tuna, minced pork. Food simulants: water; 4% acetic acid; 20%, 50% and 95% ethanol; and heptane.	GC-MS	LLE and SPE	Column: DB-1 (5 m x 0.25 mm × 0.1 μm) Injection mode: Splitless Oven temperature: 50 °C and then ramped at 20 °C/min to 300 °C.	LOQ: water and solvents: (0.5–1 μg/dm^2) Nonfatty foods (1–2 μg/dm^2) Fatty foods (2.5–5 μg/dm^2)	Migration levels from PVC stretch films into foods were 23–5,190 μg/dm^2 for DINA, 2–563 μg/dm^2 for DAA. Plasticisers migrated at high levels into lipid-soluble simulants, rapeseed oil and fatty food; and migrated in very low amounts into aqueous simulants and nonfatty food.	[132]

of ESBO into fat simulants (olive oil) was high (2,100 mg/kg of ESBO for the sample with 30% of ESBO, and even higher values for the other samples). Conversely, ESBO was not detected in the migration study performed using distilled water. In this investigation, some commercial lids were also analysed and GC–MS was used for plasticiser determination. Other plasticisers such as ATBC, dioctyl adipate and DBS were also determined in real samples. Migration analysis from three commercial lids showed that ESBO exceeded the legal migration limits in two samples, while in the other sample the migration of ATBC exceeded the legal migration limit.

Lambertini et al. [116] used LC–MS to quantify the main plasticisers used in the PVC closure gaskets for metal lids. The analysis of gasket composition showed that their formulation was mainly based on ESBO or PAD. DMP, DEP, DBS, diethyl sebacate, ATBC and 2,3-diacetoxypropyl 12-acetoxy-octadecanoate were also detected in some cases.

This study also monitored plasticiser migration into sauces placed in contact with the lids under worst-case storage conditions. In all cases, a small amount of DBS migrated into the sauce in the first 3 months, ATBC concentration did not change from the third month to the end-of-shelf life.

References

[1] G. Wypych in *Handbook of Plasticizers*, 3rd Edition, ChemTec Publishing, Toronto, Canada, 2017.

[2] J. Huang, P.N. Nkrumah, Y. Li and G. Appiah-Sefah, *Reviews of Environmental Contamination and Toxicology*, 2013, 224, 39.

[3] L.G. Krauskopf and A. Godwin in *PVC Handbook*, Eds., C.E. Wilkes, J.W. Summers and C.A. Daniels, Hanser Publishers, Munich, Germany, 2005, p.173.

[4]. H. Gallart-Ayala, O. Núñez and P. Lucci, *Trends in Analytical Chemistry*, 2013, 42, 99.

[5] S.E. Serrano, J. Braun, L. Trasande, R. Dills and S. Sathyanarayana, *Environmental Health*, 2014, 13, 1, 1.

[6]. U. Heudorf, V. Mersch-Sundermann and J. Angerer, *International Journal of Hygiene and Environmental Health*, 2007, 210, 5, 623.

[7] J.D. Meeker, S. Sathyanarayana and S.H Swan, *Philosophical Transactions of the Royal Society B: Biological Sciences*, 2009, 364, 1526, 2097.

[8] I.T. Cousins, D. Mackay and T.F. Parkerton in *Anthropogenic Compounds*, Eds., C.A. Staples, Springer, Heidelberg, Berlin, 2003, p.57.

[9] K. Mitani, F. Izushi and H. Kataoka, *Journal of Analytical Toxicology*, 2004, 28, 7, 575.

[10] H.C. Erythropel, M. Maric, J.A. Nicell, R.L. Leask and V. Yargeau, *Applied Microbiology and Biotechnology*, 2014, 98, 24, 9967.

[11] *European Union Risk Assessment Report:* Bis(2-ethylhexyl) phthalate (DEHP), CAS No: 117-81-7, 2008. http://publications.jrc.ec.europa.eu/repository/handle/JRC45705

[12] *Data on Manufacture, Import, Export, Uses and Releases of Bis(2-Ethylhexyl)Phthalate (DEHP) as Well as Information on Potential Alternatives to Its Use*, COWI A/S supported by IOM and Entec UK Ltd, ECHA/2008/02/SR5/ECA.227, 2009. https://echa.europa.eu/documents/10162/13640/tech_rep_dehp_en.pdf

[13] K.M. Rodgers, R.A. Rudel and A.C. Just in *Toxicants in Food Packaging and Household Plastics*, Ed., S.M. Snedeker, Springer-Verlag, London, UK, 2014, p.31.
[14] H.M. Koch, R. Preuss and J.D. Angerer, *International Journal of Andrology*, 2006, 29, 1, 155.
[15] K.R. Carlson and L.E. Patton in *Toxicity Review of Diisobutyl phthalate (DiBP)*, United States Consumer Product Safety Commission, 2010. https://www.cpsc.gov/s3fs-public/ToxicityReviewOfDiBP.pdf
[16] V.O. Sheftel in *Handbook of Toxic Properties of Monomers and Additives*, CRC Press, Boca Raton, FL, USA, 1995.
[17] *Diethyl Phthalate*, Concise International Chemical Assessment Document 52, World Health Organization, 2003. http://www.who.int/ipcs/publications/cicad/en/cicad52.pdf
[18] *Review of New Available Information for Dibutyl Phthalate (DBP)*, Evaluation of New Scientific Evidence Concerning the Restrictions Contained in Annex XVII to Regulation (EC) no 1907/2006 (REACH), CAS No: 84-74-2, European Chemicals Agency, 2010. https://echa.europa.eu/documents/10162/13641/dbp_echa_review_report_2010_6_en.pdf
[19] *Data on Manufacture, Import, Export, Uses and Releases of Dibutyl Phthalate (DBP) as well as Information on Potential Alternatives to its Use*, COWI A/S, Kongens Lyngby, 2009. https://echa.europa.eu/documents/10162/6ce77be0-6c61-4e95-9241-0c262817555a
[20] *Benzyl Butyl Phthalate (BBP)*, European Union Risk Assessment Report, CAS No: 85-68-7, 2007. https://echa.europa.eu/documents/10162/bad5c928-93a5-4592-a4f6-e02c5e89c299
[21] J.M. Park, M. Kim, J. Yoon, F. Kobayashi, Y. Iwasaka, C.S. Hong and Y.H. Kim, *Journal of Basic Microbiology*, 2009, 49, S31.
[22] K. Kato, M.J. Silva, C. Wolf, L.E. Gray, L.L Needham and A.M. Calafat, *Toxicology*, 2007, 236, 114.
[23] *Di-'Isononyl' Phthalate (DINP)*, Evaluation of New Scientific Evidence Concerning the Restrictions Contained in Annex XVII to Regulation (EC) no 1907/2006 (REACH), CAS No: 28553-12-0 and 68515-48-0, European Chemical Agency, 2010. https://echa.europa.eu/documents/10162/13641/dinp_echa_review_report_2010_6_en.pdf
[24] S.M. Hays, L.L. Aylward, C.R. Kirman, K. Krishnan and A. Nong, *Regulatory Toxicology and Pharmacology*, 2011, 60, 2, 181.
[25] R. Carlson in *Toxicity Review of Di-n-Octyl Phthalate (DnOP)*, United States Consumer Product Safety Commission, 2010. https://www.cpsc.gov/s3fs-public/toxicityDNOP.pdf
[26] J.H. Li, W.J. Yu, Y.H. Lai and Y.C. Ko, *The Kaohsiung Journal of Medical Sciences*, 2012, 28, 7, S10.
[27] N.H. Kleinsasser, B.C. Wallner, E.R. Kastenbauer, H. Weissacher and U.A. Harréus, *Teratogenesis, Carcinogenesis, and Mutagenesis*, 2001, 21, 3, 189.
[28] A.M. Saillenfait, J.P Sabate and F. Gallissot, *Toxicology Letters*, 2006, 165, 1, 39.
[29] M.A. Helal, *Reproductive Biology*, 2014, 14, 3, 182.
[30] *Di-butylphthalate (DBP) for Use in Food Contact Materials*, Question No: EFSA-Q-2003-192, *EFSA Journal*, 2005, 242, 1. http://www.efsa.europa.eu/sites/default/files/scientific_output/files/main_documents/242.pdf
[31] *Butyl Benzyl Phthalate (BBP) for Use in Food Contact Materials*, Question No: EFSA-Q-2003-190, *EFSA Journal*, 2005, 241, 1. http://www.efsa.europa.eu/sites/default/files/scientific_output/files/main_documents/241.pdf
[32] Bis(2-Ethylhexyl)Phthalate (DEHP) for Use in Food Contact Materials, Question No: EFSA-Q-2003-191, *EFSA Journal*, 2005, 243, 1. http://www.efsa.europa.eu/sites/default/files/scientific_output/files/main_documents/236.pdf
[33] *Di-isononyl Phthalate (DINP) for Use in Food Contact Materials*, Question No: EFSA-Q-2003-194, *EFSA Journal*, 2005, 244, 1. http://www.efsa.europa.eu/sites/default/files/scientific_output/files/main_documents/244.pdf

[34] *Di-isodecylphthalate (DIDP) for Use in Food Contact Materials*, Question No: EFSA-Q-2003-195, *EFSA Journal*, 2005, 245, 1. http://www.efsa.europa.eu/sites/default/files/scientific_output/files/main_documents/245.pdf
[35] Z. Li, L. Xu, C. Peng, H. Kuang, C. Xu, L. Wang, F. Xue, T. Ding, C. Sheng and Y. Gong, *Journal of Chromatographic Science*, 2011, 49, 4, 337.
[36] G. Cinelli, P. Avino, I. Notardonato, A. Centola and M.V. Russo, *Food Chemistry*, 2014, 146, 181.
[37] D. Xu, X. Deng, E. Fang, X. Zheng, Y. Zhou, L. Lin and Z. Huang, *Journal of Chromatography A*, 2014, 1324, 49.
[38] F. Makkliang, P. Kanatharana, P. Thavarungkul and C. Thammakhet, *Food Chemistry*, 2015, 166, 275.
[39] W.N. Wang, R.Y. Ma, Q.H. Wu, C. Wang and Z. Wang, *Talanta*, 2013, 109, 133.
[40] M.F. Zaater, Y.R. Tahboub and A.N. Al Sayyed, *Journal of Chromatographic Science*, 2014, 52, 5, 447.
[41] D.S. Milojković, D.H. Anđelković, G.M. Kocić and T.D. Anđelković, *Journal of the Serbian Chemical Society*, 2015, 80, 8, 983.
[42] M. Asensio-Ramos, L.M. Ravelo-Pérez, M.Á. González-Curbelo and J. Hernandez-Borges, *Journal of Chromatography A*, 2011, 1218, 42, 7415.
[43] H. Farahani, M.R. Ganjali, R. Dinarvand and P. Norouzi, Talanta, 2008, 76, 4, 718.
[44] M. Li, Y. Li and L. Mao, *Chinese Journal of Chromatography*, 2007, 25, 1, 35.
[45] L.P. Zhu, T. Zhu, Y.P. Ma, Y.F. Ni, Y. Wang and Q.C. Yan, *Chinese Journal of Analytical Chemistry*, 2013, 41, 7, 1019.
[46] M.A. Farajzadeh, D. Djozan, M. Reza, A. Mogaddam and J. Norouzi, *Journal of Separation Science*, 2012, 35, 5–6, 742.
[47] M. Rezaee, Y. Assadi, M.R.M. Hosseini, E. Aghaee, F. Ahmadi and S. Berijani, *Journal of Chromatography A*, 2006, 1116, 1, 1.
[48]. H. Yan, X. Cheng and B. Liu, *Journal of Chromatography B*, 2011, 879, 25, 2507.
[49] Y. Picó, M. Fernández, M.J. Ruiz and G. Font, *Journal of Biochemical and Biophysical Methods*, 2007, 70, 2, 117.
[50] A. Mohamed and A. Ammar, *American Journal of Food Technology*, 2008, 3, 5, 341.
[51] Y. Liu, S. Wang and L. Wang, *Journal of Agricultural and Food Chemistry*, 2013, 61, 6, 1160.
[52] J.D. Carrillo, M.P. Martínez and M.T. Tena, *Journal of Chromatography A*, 2008, 1181, 1–2, 125.
[53] A. Spietelun, L. Marcinkowski, M. de la Guardia and J. Namieśnik, *Journal of Chromatography A*, 2013, 1321, 1.
[54] H. Chen, X.J. Liu, C. Yang, J. Gao, C.W. Ye and X.J. Li, *Chromatographia*, 2009, 70, 5–6, 883.
[55] J. Meng J. Bu, C. Deng and X. Zhang, *Journal of Chromatography A*, 2011, 1218, 12, 1585.
[56]. I.S. Ibarra, J.A. Rodriguez, C.A. Galán-Vidal, A. Cepeda and J.M. Miranda, *Journal of Chemistry*, 2015, 1.
[57] L.X. Yi, R. Fang and G.H. Chen, *Journal of Chromatographic Science*, 2013, 51, 7, 608.
[58] J. He, R. Lv, J. Zhu and K. Lu, *Analytica Chimica Acta*, 2010, 661, 2, 215.
[59] M. Anastassiades, S.J. Lehotay, D. Štajnbaher and F.J. Schenck, *Journal of AOAC International*, 2003, 86, 2, 412.
[60] P. Yin, X. Liu, H. Chen, R. Pan and G. Ma, *Food Additives & Contaminants: Part A*, 2014, 31, 8, 1406.
[61] J. Regueiro, M. Llompart, C. Garcia-Jares, J.C. Garcia-Monteagudo and R. Cela, *Journal of Chromatography A*, 2008, 1190, 1, 27.
[62] H. Yan, X. Cheng and K. Yan, *Analyst*, 2012, **137**, 20, 4860.
[63] A. Prieto, O. Basauri, R. Rodil, A. Usobiaga, L.A. Fernández, N. Etxebarria and O. Zuloaga, *Journal of Chromatography A*, 2010, 1217,16, 2642.

[64] J.I. Cacho, N. Campillo, P. Vinas and M. Hernandez-Cordoba, *Journal of Chromatography A*, 2012, 1241, 21.

[65] I. Ostrovský, R. Čabala, R. Kubinec, R. Górová, J. Blaško, J. Kubincová and W. Lorenz, *Food Chemistry*, 2011, 124, 1, 392.

[66] P. Wu, D. Yang, L. Zhang, X. Shen, X. Pan, L. Wang and Y. Ying, *Journal of Separation Science*, 2012, 35, 21, 2932.

[67] B. Cavaliere, B. Macchione, G. Sindona and A. Tagarelli, *Journal of Chromatography A*, 2008, 1205, 1, 137.

[68] M. Sun, R. Tang, Q. Wu, C. Wang and Z. Wang, *Analytical Methods*, 2013, **5**, 20, 5694.

[69] G.K. Mortensen, K.M. Main, A.M. Andersson, H. Leffers and N.E. Skakkebæk, *Analytical and Bioanalytical Chemistry*, 2005, 382, 1084.

[70] G. Cinelli, P. Avino, I. Notardonato, A. Centola and M.V. Russo, *Analytica Chimica Acta*, 2013, 769, 72.

[71] M.V. Russo, P. Avino and I. Notardonato, *Journal of Chromatography A*, 2016, 1474, 1.

[72] R.L. Self and W.H. Wu, *Food Control*, 2012, 25, 1, 13.

[73] L. Ackerman, G. Noonan and T. Begley, *Food Additives & Contaminants*, 2009, 26, 12, 1611.

[74] J. Guo, K. Luo, D. Chen, X. Tan and Z. Song, *Journal of Food Composition and Analysis*, 2013, 31, 2, 226.

[75] G. Chen, H. Hu, T. Wu, P. Tong, B. Liu, B. Zhu and Y. Du, *Food Control*, 2014, 35, 1, 218.

[76] J. Yang, Y. Li, Y. Wang, J. Ruan, J. Zhang and C. Sun, *TrAC Trends in Analytical Chemistry*, 2015, 72, 10.

[77] *RASFF – Food and Feed Safety Alerts*, Rapid Alert System for Food and Feed. https://ec.europa.eu/food/safety/rasff_en

[78] Commission European Regulation (EU) No 10/2011 of 14th January 2011 on Plastic Materials and Articles Intended to Come into Contact with Food, *Official Journal of the European Union L*, 2011, 12, 1.

[79] L. Coltro, J.B. Pitta, P.A. da Costa, M.Â.F. Perez, V.A. de Araujo and R. Rodrigues, *Food Control*, 2014, 44, 118.

[80] T. Fierens, G. Vanermen, M. van Holderbeke, S. De Henauw and I. Sioen, *Food and Chemical Toxicology*, 2010, 50, 12, 4428.

[81] T. Fierens, K. Servaes, M. van Holderbeke, L. Geerts, S. De Henauw, I. Sioen and G. Vanermen, *Food and Chemical Toxicology*, 2012, 50, 7, 2575.

[82] E.L. Bradley, R.A. Burden, I. Leon, D.N. Mortimer, D.R. Speck and L. Castle, *Food Additives & Contaminants: Part A*, 2013, 30, 4, 722.

[83] X.L. Cao, W. Zhao and R. Dabeka, *Food Additives & Contaminants: Part A*, 2015, 32, 11, 1893.

[84] P. Serodio and J.M.F. Nogueira, *Water Research*, 2006, 40, 13, 2572.

[85] K. Zhang, G.O. Noonan and T.H. Begley, *Food Additives & Contaminants*, 2008, 25, 11, 1416.

[86] A. Sannino, *Food Additives & Contaminants*, 2009, 2, 2, 166.

[87] J.J. Ríos, A. Morales and G. Marquez-Ruiz, *Talanta*, 80, 2010, 2076.

[88] Y. Guo, Z. Zhang, L. Liu, Y. Li, N. Ren and K. Kannan, *Journal of Agricultural and Food Chemistry*, 2012, 60, 27, 6913.

[89] A. Schecter, M. Lorber, Y. Guo, Q. Wu, S.H. Yun, K. Kannan, M. Hommel, N. Imran, L. Hynan, D. Cheng, J.A. Colacino and L.S. Birnbaum, *Environmental Health Perspectives*, 2013, 121, 4, 473.

[90] S. Keresztes, E. Tatár, Z. Czégény, G. Záray and V.G. Mihucz, *Science of the Total Environment*, 2013, 458, 451.

[91] L.K Sørensen, *Rapid Communications in Mass Spectrometry*, 2006, 20, 7, 1135.

[92] Y. Ma, Y. Hashi, F. Ji and J.M. Lin, *Journal of Separation Science*, 2010, 33, 2, 251.

[93] Q. Xu, X. Yin, M. Wang, H. Wang, N. Zhang, Y. Shen, S. Xu, L. Zhang and Z. Gu, *Journal of Agricultural and Food Chemistry*, 2010, 58, 21, 11311.
[94] Z. Guo, D. Wei, M. Wang and S. Wang, *Journal of Chromatographic Science*, 2010, 48, 9, 760.
[95] A. Khedr, *Journal of Chromatography B*, 2013, 930, 98.
[96] Y. Hayasaka, *Journal of Chromatography A*, 2014, 1372, 120.
[97] N. Ibrahim, R. Osman, A. Abdullah and N. Saim, *Journal of Chemistry*, 2014, 2014, 1.
[98] M.A. Moreira, L.C. André and Z. de Lourdes Cardeal, *Food Chemistry*, 2015, 178, 195.
[99] J. Pérez-Outeiral, E. Millán and R. Garcia-Arrona, *Food Control*, 2016, 62, 171.
[100] S. Gärtner, M. Balski, M. Koch and I. Nehls, *Journal of Agricultural and Food Chemistry*, 2009, 57, 22, 10675.
[101] X. Gao, B. Yang, Z. Tang, X. Luo, F. Wang, H. Xu and X. Cai, *Journal of Chromatographic Science*, 2014, 52, 5, 383.
[102] A. Fankhauser-Noti and K. Grob, *Analytica Chimica Acta*, 2007, 582, 2, 353.
[103] A.D. Godwin in *Applied Plastics Engineering Handbook: Processing and Materials*, 1st Edition, Ed., M. Kutz, Elsevier, Atlanta, GA, USA, 2011, p.487.
[104] H.D. Barros, H.P.D.S. Zamith, F.S. Bazílio, L.J.D. Carvalho and S.D.M.P. Abrantes, *Food Science and Technology (Campinas)*, 2011, 31, 2, 547.
[105] *A Total Reduction Factor of 5 for Di(2-Ethylhexyl) Adipate used as Plasticizer in Flexible PVC Food Packaging Films*, Question No: EFSA-Q-2003-071, *EFSA Journal*, 2005, 217, 1. http://www.efsa.europa.eu/sites/default/files/scientific_output/files/main_documents/217.pdf
[106] V. Lajqi-Makolli, M. Kerolli-Mustafa, I. Malollari and J. Lajqi, *Materialwissenschaft und Werkstofftechnik*, 2015, 46, 1, 16.
[107] A. Fankhauser-Noti, S. Biedermann-Brem and K. Grob, *European Food Research and Technology*, 2006, 223, 4, 447.
[108] M.J. Silva, E. Samandar, A.M. Calafat and X. Ye, *Toxicology In Vitro*, 2015, 29, 4, 716.
[109] R. Nagorka, A. Conrad, C. Scheller, B. Sussenbach and H.J. Moriske, *International Journal of Hygiene and Environmental Health*, 2011, 214, 1, 26.
[110] A. Wypych in *Plasticizers Databook*, ChemTec Publishing, Toronto, Canada, 2013.
[111] T.T. Bui, G. Giovanoulis, A.P. Cousins, J. Magnér, I.T. Cousins and C.A. de Wit, *Science of the Total Environment*, 2016, 541, 451.
[112] V.O Sheftel in *Indirect Food Additives and Polymers: Migration and Toxicology*, Lewis Publishers, Boca Raton, FL, USA and London, UK, 2000.
[113] M. Rahman and C.S. Brazel, *Progress in Polymer Science*, 2004, 29, 12, 1223.
[114] *Eastman 168™ Non-Phthalate Plasticizer for Vinyl Plastisols and Vinyl Compounds*, Eastman, Kingsport, TN, USA, 2011. http://www.eastman.com/Literature_Center/P/PZ128.pdf
[115] J.H. Petersen and T. Breindahl, *Food Additives & Contaminants*, 2000, 17, 2, 133.
[116] F. Lambertini, D. Catellani, M. Vindigni and M. Suman, *Journal of Mass Spectrometry*, 2016, 51, 9, 805.
[117] F.C.Y. Wang, *Journal of Chromatography A*, 2000, 883, 199.
[118] Fankhauser-Noti, K. Fiselier, S. Biedermann-Brem and K. Grob, *Journal of Chromatography A*, 2005, 1082, 2, 214.
[119] A.E. Goulas, K.I. Anifantaki, D.G. Kolioulis and M.G. Kontominas, *Journal of Dairy Science*, 2000, 83, 8, 1712.
[120] E. Goulas, P. Zygoura, A. Karatapanis, D. Georgantelis and M.G. Kontominas, *Food and Chemical Toxicology*, 2007, 45, 4, 585.
[121] Z. Ežerskis, V. Morkūnas, M. Suman and C. Simoneau, *Analytica Chimica Acta*, 2007, 604, 1, 29.
[122] M. Suman, E. De Dominicis and I. Commissati, *Journal of Mass Spectrometry*, 2010, 45, 9, 996.
[123] Y. Tsumura, S. Ishimitsu, A. Kaihara, K. Yoshii and Y. Tonogai, *Journal of Health Science*, 2002, 48, 6, 493.

[124] K. Hirayama, H. Tanaka, K. Kawana and H. Nakazawa, *Food Additives & Contaminants*, 2001, 18, 4, 357.
[125] P. Kueseng, P. Thavarungkul and P. Kanatharana, *Journal of Environmental Science and Health B*, 2007, 42, 569.
[126] X.L. Cao, *Journal of Chromatography A*, 2008, 1178, 1, 231.
[127] G.A. Pedersen, L.K. Jensen, A. Fankhauser, S. Biedermann, J.H Petersen and B. Fabech, *Food Additives & Contaminants*, 2008, 25, 4, 503.
[128] A. Sannino, *Journal of AOAC International*, 2010, 93, 1, 315.
[129] X.L. Cao, W. Zhao, R. Churchill and R. Dabeka, *Journal of Food Protection*, 2013, 76, 11, 1985.
[130] X.L. Cao, W. Zhao, R. Churchill and C. Hilts, *Journal of Food Protection*, 2014, 77, 4, 610.
[131] L. Carnol, C. Schummer and G. Moris, *Food Analytical Methods*, 2017, 10, 2, 298.
[132] Y. Kawamura, Y. Ogawa and M. Mutsuga, *Food Science & Nutrition*, 2017, 5, 3, 390.
[133] P.D. Zygoura, A.E. Goulas, K.A. Riganakos and M.G. Kontominas, *Journal of Food Engineering*, 2007, 78, 3, 870.
[134] C. Bueno-Ferrer, A. Jiménez and M.C. Garrigós, *Food Additives & Contaminants*, 2010, 27, 10, 1469.

5 Photoinitiators

5.1 General aspects of photoinitiators

Photoinitiators play an important role in the curing process of polymeric materials. These compounds absorb ultraviolet (UV) light and generate reactive species (free radicals or ions, usually cations), which are capable of initiating the curing process (polymerisation) [1].

Photoinitiators are an alternative to thermally decomposing initiators (thermal polymerisation process), which offer the advantage of defining the exact start and end points of the polymerisation process via the duration of the irradiation period, as the process of photoinitiator decomposition is strongly dependent on the UV light intensity. In addition, the photocuring process offers a high rate of polymerisation at low-temperature operation, shortened cure times, lower energy requirements, solvent-free formulation and reduction of volatile organic compound emissions [2].

Because of the advantages of this process, photopolymerisation is widely used in several applications such as inks, adhesives, coatings, varnishes, photolithography, dental materials, electronics, orthopaedic biomaterials, cosmetics and many others. The selection of a photoinitiator is dependent on the application. UV-curing printing inks and coatings are widely used in food-packaging applications and are normally applied to the exterior of flexible food packaging or paper and board secondary food packaging. UV curable varnishes are widely used on food cartons and labels [3]. In addition to photoinitiators, various additives such as light stabilisers, dyes and fillers can be added in order to improve the final properties of the photopolymer.

During the photopolymerisation process, monomers combine to create a polymer upon irradiation with UV light. There are two main types of photoinitiators: free radical and cationic photoinitiators, that is, a photoinitiating system creates free radicals or cations under UV radiation, which react with monomers and/or oligomers. In general, the polymerisation process comprises four main steps: initiation by active centres, propagation, chain transfer and termination [4].

Photoinitiated radical polymerisation is most commonly used for industrial applications, in comparison to cationic photopolymerisation. A variety of free radical photopolymerisation materials have been commercialised, including methacrylate monomers and oligomers. Free radical photopolymerisation systems offer many advantages, including higher polymerisation rates, short-cure time processes in the absence of a solvent under ambient conditions and flexible monomer chemistry, among others. However, oxygen is an inhibitor of free radical photopolymerisation and several methods have been applied to mitigate oxygen inhibition, including the use of Type II photoinitiators, increasing UV radiation intensity, curing in an

https://doi.org/10.1515/9783110648065-005

inert atmosphere, addition of oxygen scavengers, the use of dye sensitisers, the use of surface active initiators and the use of a high photoinitiator concentration [5]. Free radical photopolymerisation involves the generation of active species (free radicals) through the absorption of UV or visible light (initiation). During propagation, oligomers and monomers add to the growing polymer chain (propagation), which eventually stops growing and begins the formation of a new polymer chain (chain transfer) and finally, polymerisation stops when two growing polymer chains react with each other (termination) [5].

There are two main types of free radical photoinitiators, that is, Type I and Type II. The reaction mechanisms of free radical photoinitiators are well established. Types I and II photoinitiators have different mechanisms by which initiating radicals are formed [4, 6]:

- Type I photoinitiators (direct cleavage): after light absorption, the photoinitiator undergoes unimolecular bond cleavage (usually α-cleavage) to generate free radicals.
- Type II photoinitiators (bimolecular reaction): after light absorption, the photoinitiator (excited state) reacts with a coinitiator or synergist to generate free radicals.

Table 5.1 lists the main families of Types I and II photoinitiators.

Cationic photopolymerisation has undergone a slower development compared with that of free radical photopolymerisation. However, in recent years, there has been significant progress in the application of these photoinitiators due to their great potential and several advantages. Contrary to free radical photopolymerisation, cationic processes are not inhibited by oxygen and the initiating species are Brønsted or Lewis acids (usually chemically stable for a long time); however, cationic photopolymerisation may be inhibited by bases and water (acting as a chain transfer agent). This technique can be used to polymerise monomers that cannot be polymerised by free radicals (vinyl ethers, epoxides, siloxanes, propenyl ether, cyclic sulfides, lactones and lactams, among others). Most widely used are diaryliodonium salts (absorption at 220–270 nm) and triarylsulfonium salts (strong absorption at 220–230 nm and weak around 300–325 nm) [4, 6, 8]. During this type of photopolymerisation, the cationic photoinitiator forms a cation, as a result of exposure to UV radiation or visible light. The cation initiates the polymerisation process and termination is usually controlled by a diffusion mechanism. Some examples of free radical and cationic photoinitiators are listed in Table 5.2.

The European Union (EU) does not have specific regulations for printing inks for food contact materials (FCM). However, Specific migration limit (SML) for some photoinitiators are established in European Commission Regulation 10/2011 on plastic materials and articles intended to come into contact with food [9]. An EU Member State developed laws relating to inks for food packaging; the Swiss authorities drew up an ordinance for materials and articles in contact with food (SR 817.023.21). This document

Table 5.1: Main families of Types I and II photoinitiators [6, 7].

Free radical photopolimerization			
Type I	**Absorption UV range**	**Applications**	**References**
Hydroxyacetophenones	Between 230 and 270 nm and weaker up to 360 nm	Mainly used as precursors for resins and fragrances as well as in clear coatings, overprint varnishes, topcoats for wood, metal, plastics, adhesives among others.	[6, 7]
Alkylaminoacetophenones (AAAPs)	280–350 nm	Used in high-speed offset and flexo inks, UV ink-jet, etch resists, printing plates and solder masks.	[6]
Benzylketals	Between 230 and 270 nm	Used to improve the shelf life of formulations. Important alpha cleavage photoinitiators in UV curing industry.	[6, 7]
Benzoin ethers	230–270 nm	Efficient in unsaturated polyester/styrene formulations. Low-cost materials for wood coatings and putties for particleboard.	[6]
Phosphine oxides	Between 350 and 420 nm	Used for thick film, highly pigmented formulations and for white lacquers containing titanium dioxide. Also used in printing plates and glass fiber-reinforced polyesters.	[6]
Type II			
Benzophenones	Between 230 and 260 nm	Widely used in varnishes and inks. Benzophenone is also used in perfumery.	[6]
Substituted benzophenones	Between 280 and 330 nm	Benzophenones substitutes such as 4-phenylbenzophenone, PBZ, Trigonal 12 (58) are considerably more reactive than BP and is used in inks to add some reactivity.	[6]

(continued)

Table 5.1 (continued)

Free radical photopolimerization			
Type I	**Absorption UV range**	**Applications**	**References**
Thioxanthones	Between 350 and 410 nm	Used almost universally in inks formulations. Very useful in pigmented coatings. Other applications include food-packaging materials.	[6]
Anthraquinones	280–320 nm	Used for UV-curable printing plates in the early days of UV curing.	[6]
Benzylformate esters	230–260 nm	Clear coatings, base coat for plastics and metals, among others.	[6]
Camphorquinone	468 nm	Mainly used in dental applications.	[6]

Table 5.2: Examples of free radical photoinitiators (Types I and II), amine synergist and cationic photoinitiators.

Structure	Acronym	Common name	Type	CAS	MW	Melting point (°C)	Boiling point (°C)	LOG P (o/w)	Vapour prssure (mm Hg)	SML (mg/kg)
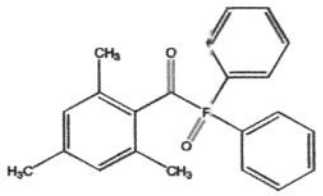	TPO	Trimethylbenzoyl diphenylphosphine oxide	I	75980-60-8	348.37	NA	519.6 ± 60.0**	4.620 ± 0.420**	6.71E-11**	0.05
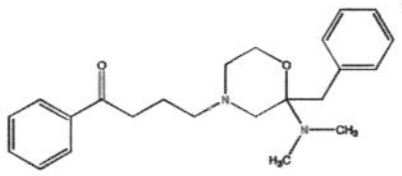	BDMB	2-Benzyl-2-dimethylamino-4-morpholino butyrophenone	I	119313-12-1	366.50	NA	528.8 ± 50.0**	3.38 ± 0.469**	2.87E-11**	–
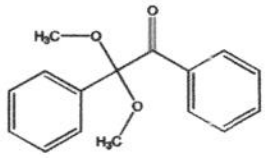	DMPA	2,2-Dimethoxy-2-phenyl acetophenone	I	24650-42-8	256.30	63–64 *	100–115 *	3.62 ± 0.486**	1.06E-5**	–
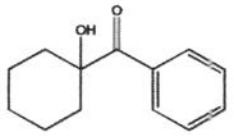	HCPK	1-Hidroxycyclohexyl-phenyl-ketone	I	947-19-3	204.26	47–49 *	140 *	2.175 ± 0.276** (25 °C)	3.68E-5 **	–
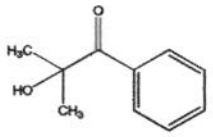	HMPP	2-Hidroxycy-2-methyl propiophenone	I	7473-98-5	164.20	>186 *	116–118 *	1.485 ± 0.380** (25 °C)	6.08E-3**	–

(continued)

Table 5.2 (continued)

Structure	Acronym	Common name	Type	CAS	MW	Melting point (°C)	Boiling point (°C)	LOG P (o/w)	Vapour prssure (mm Hg)	SML (mg/kg)
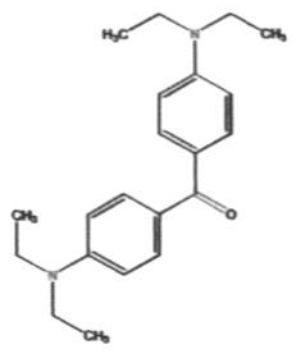	DEAB	Benzophenone, 4,4-bis(diethylamino)-	II	90-93-7	324.46	158 *	475.7 ± 30.0**	5.908 ± 0.354 ** (25 °C)	3.25E-9**	–
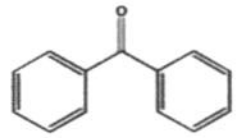	BP	Benzophenone	II	119-61-9	182.22	48.5 *	305.4 *	3.21 **	8.23E-4**	0.6*
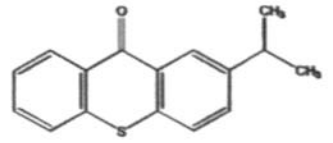	ITX	2-Isopropyl thioxanthone	II	5495-84-1	254.35	77.5 *	398.9 ± 32.0 **	5.113 ± 0.214**	1.43E-6**	0.05
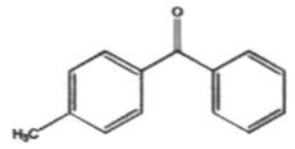	4-MBP	Benzophenone, 4-methyl-	II	134-84-9	196.24	59.5 *	164–169 *	3.874 ± 0.306**	1.94E-4**	0.6*
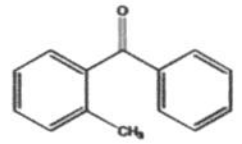	2-MBP	Benzophenone, 2-methyl-	II	131-58-8	196.24	< -18*	308*	3.784 ± 0.317**	6.17E-4**	0.6*

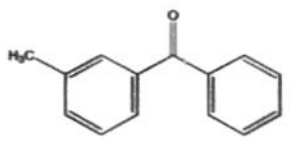	3-MBP	Benzophenone, 3-methyl-	II	643-65-2	196.24	159–160*	317*	3.857 ± 0.308**	3.92E-4**	0.6*
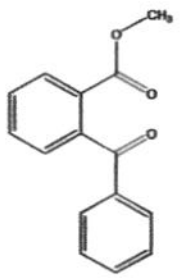	MBB	Methyl-2-benzoylbenzoate	II	606-28-0	240.25	52*	351*	2.828 ±0.315**	4.08E-5 **	0.05
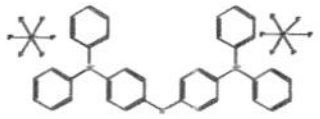	BIS	Bis (4-(diphenylsulfonio) phenyl)sulfide bis (hexafluorophosphate)	Cationic	74227-35-3	N/A	N/A	N/A	N/A	N/A	–
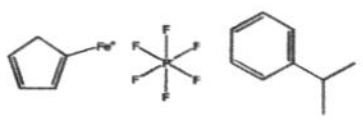	Irg 261	(Cumene) cyclopentadienyliron(II) hexafluorophosphate	Cationic	32760-80-8	400.12	N/A	N/A	N/A	N/A	–
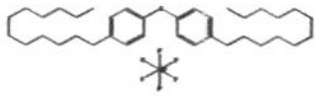	–	Bis(4-dodecylphenyl) iodonium hexafluroantimonate	Cationic	71786-70-4	N/A	N/A	N/A	N/A	N/A	–
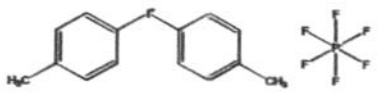	–	Bis(4-ethylphenyl) iodonium hexafluorophosphate	Cationic	60565-88-0	N/A	N/A	N/A	N/A	N/A	–
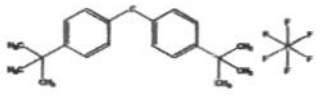	–	Bis(4-tert-butylphenyl) iodonium hexafluorophosphate	Cationic	0061358-25-6	N/A	N/A	N/A	N/A	N/A	–

(continued)

Table 5.2 (continued)

Structure	Acronym	Common name	Type	CAS	MW	Melting point (°C)	Boiling point (°C)	LOG P (o/w)	Vapour prssure (mm Hg)	SML (mg/kg)
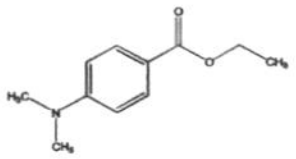	–	Ethyl-4-(dimethylamino)-benzoate	Amine Synergist	10287-53-3	193.24	65–66*	125–142*	2.511 ± 0.236**	1.43E-3**	–
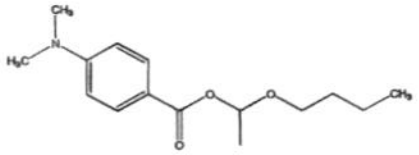	–	Butoxyethyl-4-(dimethylamino) benzoate	Amine Synergist	67362-76-9	265.35	NA	357.9 ± 22.0**	3.414 ± 0.501**	2.65E-5**	–
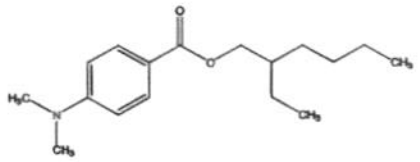	–	2-Ethylhexyl-4-dimethylamino benzoate	Amine Synergist	21245-02-3	277.40	242.5–243.5*	382.9 ± 25.0**	5.412 ± 0.241**	4.57E-6 **	–

* Experimental and ** predicted data. Obtained from SciFinder® version web. N/A: not available.

details certain provisions relating to packaging inks and permitted substances, which are listed in Annex 1 (lists I, II and III) and in list V of Annex 6 (Part A: evaluated substances and Part B: nonevaluated substances) [10].

On the other hand, in 2013, the European Printing Inks Association published a list of *Photoinitiators for Low Migration UV Printing Inks and Varnishes* that contains 123 photoinitiators classed into two large groups: group 1 (for all packaging types) and group 2 (only for packaging such as some metal packaging where transfer/migration of <10 ppb or other required limits can be achieved) [11].

5.2 Migration of photoinitiators

The UV-curing inks that are used in food applications usually have complex compositions, and basically contain a mixture of oligomers, monomers, pigments and photoactive compounds, called *photoinitiators*. Printing inks are commonly used on the external side of FCM, or between one of the outer layers of a multilayer material.

The printing process involves the use of several techniques, which are also applicable to food-packaging materials. The most commonly used technologies for printing include offset, flexography, gravure printing and ink-jet printing. Of these techniques, offset printing and flexographic printing are the most frequently used methods for printing paper and cardboard FCM [12].

Components from the printing inks can migrate into packaged foodstuffs; in addition, the migration of 'nonintentionally added substances' derived from printing inks can also occur. These can be reaction products of ink components or created by the reaction of inks with other packaging components [13].

The migration of photoinitiators can occur in different ways: 1) the direct transfer of compounds from the printed surface of the food packaging via diffusion, that is, low MW compounds may diffuse through the food contact packaging and migrate into the foodstuff, 2) 'set-off' migration refers to the transfer of chemicals from the printed side of a packaging material to the food contact surface (unprinted side). Migration 'set-off' normally occurs when stacking or rolling packaging materials prior to packaging food products and 3) indirect transfer via the vapour phase can also occur [13–15].

In Europe, notifications regarding the migration of photoinitiators from FCM into food are made through the Rapid Alert System for Food and Feed (RASFF). This system was created with the aim of providing a rapid exchange of information between EU Member States with relation to national controls on food and feed products.

In September 2005, an RASFF notified case of public concern involved the Italian authorities detecting a photoinitiator, isopropylthioxanthone (ITX), in baby milk from Spain. This compound was a component of the UV-cured ink used on the packaging. Contamination of ITX in milk products packed in Tetrapack is probably the result of the set-off phenomenon, because aluminium layers do not allow ink component migration through packaging material. In December 2005, ITX was also found in olive oil, wine and fruit juices [15, 16].

Between 2000 and 2011, more than 100 alerts relating to the migration of photoinitiators have been reported by the RASFF system. In 2012 and 2013, no notifications involving photoinitiators were reported; however, since 2014, four notifications have been reported [BP migration (3 notifications) and BP + 2,2-dimethoxy-2-phenylacetophenone (1 notification)] [17, 18].

The migration of photoinitiators is a concern to public health. In fact, the toxicity of BP, a photoinitiator widely used in several applications, has been investigated, with a focus on certain aspects including genotoxicity, carcinogenicity, as well as oestrogenic and antiandrogenic activities. The results of experimental studies showed that BP has no genotoxic potential; however, damage to the liver and kidney has been reported in carcinogenicity studies [19].

On the other hand, results of the toxicological evaluation of ITX and 2-ethylhexyl-4-dimethylaminobenzoate (EHDAB) in FCM showed no genotoxic potential for ITX and EHDAB; the latter is also not teratogenic. Exposure to ITX and EHDAB can occur from food and beverages packaged in cartons printed with UV offset inks [20].

Other more recent studies of thioxanthone derivatives (2-ITX, 4-isopropylthioxanthone (4-ITX), 2,4-diethylthio xanthone, 2-chlorothioxanthone and 1-chloro-4-propoxythioxanthone) have shown endocrine-disrupting activities at the level of nuclear receptor signalling and steroid hormone production [21].

Driven by the toxicological concern regarding photoinitiator migration, several analytical techniques have been developed for the analysis of photoinitiators and their by-products in packaging materials, food and food simulants. Gas chromatography (GC) and liquid chromatography (LC) techniques are often applied to study the migration of certain photoinitiators from printed food-packaging materials. Of these, LC coupled to diode array detection (DAD), fluorescence detection (FLD), mass spectrometry (MS) or tandem mass spectrometry (MS/MS) is the most commonly used analytical technique for the determination and quantification of photoinitiators in FCM and food products.

GC combined with MS or MS/MS and flame ionisation detection (FID) are also employed for the determination of photoinitiators in food products.

Jung and Simat [22] developed a method involving high-performance liquid chromatography (HPLC) with DAD method for the screening of 63 monomeric and polymeric photoinitiators and amine synergists in eight food-packaging materials (plastic, paper, cartonboard) printed with printing inks and/or varnishes. Detection limits ranged between 0.02 and 5.5 $\mu g/dm^2$. Results showed that the method was suitable for the determination of photoinitiators and amine synergists in food-packaging materials as well as for the analysis of food simulants.

Sun et al. [23] developed a method based on LC–MS/MS, suitable for detecting low levels of ITX in different matrices (acidic juices to fatty milk and yoghurt drink samples). The detection limit was 0.15 μg/kg and recoveries ranged from 97% to 103%. ITX was detected in 7 of the 39 food samples at concentrations between 0.53 and 84.30 μg/kg.

In 2010, Koivikko et al. [24] used HPLC–DAD to determine BP, 4-MBP and related derivatives [benzophenone acrylate (BPAcr) and PBZ] in paperboard food packaging. Chromatographic analyses were performed using an Eclipse XDB–C_{18} column (250 mm × 4.6 mm × 5 μm) and acetonitrile (ACN) and water as the mobile phase, employing a gradient mode. The most abundant compound found in tested samples was BP (59% of the samples) followed by 4-MBP (30%), BPAcr (9%) and PBZ (7%). The identification of compounds was confirmed with GC–MS.

Bradley et al. [25] used GC–MS to detect 20 printing ink compounds in 350 foodstuffs (snacks, sweets, chocolates, biscuits and crisps, bakery products, baking ingredients, cereals, frozen products, rice and pasta, desserts, gravy/cooking sauces, teabags, fruit juice, sandwiches and other foods) packaged in printed paper/board. Chromatographic separation was achieved with a ZB-5ms (30 m × 0.25 mm × 0.25 μm) column and the oven temperature was programmed to increase from 100 to 320 °C. The injection was in splitless mode. The migration of several printing ink compounds, such as BP, 4-MBP, MBB, 1-hydroxycyclohexyl-phenyl-ketone (HCPK), ethyl-4-dimethylaminobenzoate (EDMAB) and EHDAB, was detected. BP was confirmed to be present in 37 foodstuffs in concentrations ranging from 0.01 to 2.46 mg/kg.

Table 5.3 lists GC and LC methods reported for the analysis of photoinitiators in foodstuffs, packaging materials and food simulants.

Others authors have used high-performance thin layer chromatography (HPTLC) coupled to FLD, ESI–MS and DART–MS. HPTLC is a simple, robust and effective technique in the quantitative analysis of compounds [47]. Over the last few years, a technique called DART has also been used and its advantages include that it does not require sample treatment; however, this technique is not highly quantitative [48]. Micellar electrokinetic chromatography (MEKC) with DAD was recently applied with successful results for the determination of photoinitiators in fruit juice [49].

For the analysis of photoinitiators in food matrices, extraction and clean-up procedures are generally required to avoid matrix interference. Several methods have been reported in the literature for photoinitiator extraction from food products, including LLE, SPE, SPME, pressurised liquid extraction (PLE) and QuEChERS.

LLE, using ACN or hexane as the extractant, is frequently used for the extraction of photoinitiators from food matrices, especially from liquid and fatty samples. In 2008, Benetti et al. [37] published a method based on the ACN extraction of ITX from dairy products, employing LC/MS. In the same year, Sanchez Silva et al. [50] used ammoniac and hexane to extract six photoinitiators [1-hydroxycyclohexyl-1-phenyl ketone (Irgacure® 184), benzophenone, ethanone, 2,2-dimethoxy-1,2-diphenyl- (9CI) (Irgacure® 651), 1-propanone, MTMP (Irgacure® 907), quantacure ITX and quantacure

Table 5.3: GC and LC methods for photoinitiators determination in food products, food simulants and FCMs [14, 15, 26–46].

Analytes	Matrix	Detector	Sample preparation or extraction	Column/injection mode/oven temperature	LOD/(LOQ)	Results	References
GC methods							
Benzophenone (BP)	Food: 350 samples of food FCM: cartonboard	MS	LLE	Column: RTX-1 (60 m × 0.25 mm; 0.25 μm) Injection mode: Splitless Oven temperature: 50 °C (1 min) and then raised at 20 °C min to 280 °C (5 min).	0.01 mg/kg /(0.05 mg/kg (in food)	BP was detected in 51 of the 71 food samples. The levels of BP in food were in a range of 0.01 to 7.3 mg/kg.	[26]
Isopropyl thioxanthenone (ITX)	Food: milk	MS/MS	Solid-phase extraction (SPE)	Column: Restek RTX®-5MS (30 m × 0.25 mm ID × 0.25 μm column) Injection mode: Splitless Oven temperature: 100 °C (1 min), increased to 180 °C at rate 80 °C /min, then increased to 280 °C at rate 25 °C/min and finally kept at 280 °C (5 min).	0.1 μg/L / (0.5 μg/L)	Detectable amounts of ITX were found in all milk samples (4.0–53.0 μg/L).	[27]

1-Hydroxycyclohexyl -1-phenylketone (Irgacure 184), 2,2-dimethoxy-2-phenylacetophenone (Irgacure-651).	Food simulant: 10% ethanol (v/v), 95% ethanol (v/v), isooctane	FID	LLE for 10% ethanol and 95% ethanol. Isooctane: concentration	Column: HP-5 (30 m × 0.32 mm × 0.25 μm) Injection mode: Splitless Oven temperature: 100 °C, then increased by 5 °C/min to up to 200 °C (2 min), then increased by 15 °C/min to up to 300 °C (5 min).	–	Irgacure-184 was shown to be quite stable under heat exposure conditions tested and in all food simulants. Irgacure-651 was stable in all food simulants, except in 10% ethanol (v/v) at 60 °C.	[28]
BP	FCM: cartonboard Food: cornish pie, cheese, egg and bacon flan, breaded fish sticks, potato waffles, cheese and onion sticks	MS	Extraction with solvent. clean up: size exclusion chromatographic	–	–	In cartonboard BP was detected in four of the seven products (0.4–3.0 mg/dm^2). BP was detected in pie, flan and cheese and onion sticks at concentrations of 0.6–2.9 mg/kg. Four of the food samples did not contain BP above 0.05 mg/kg food.	[29]

(continued)

Table 5.3 (continued)

GC methods							
Analytes	**Matrix**	**Detector**	**Sample preparation or extraction**	**Column/injection mode/oven temperature**	**LOD/(LOQ)**	**Results**	**References**
BP and 4-methylben zophenone (4MBP)	Food: breakfast cereals	MS	LLE followed by purification by SPE	Column: SGE BPX-5 (25 m × 0.22 mm × 0.25 μm) Injection mode: Splitless Oven temperature: 50 °C (1 min) to 280 °C at rate of 25 °C/min (5 min)	BP: 2 μg/kg / (6 μg/kg) 4MBP: 2 μg/kg / (8 μg/kg)	The method was validated in terms of recovery (between 74% and 98%), repeatability (CVr = 2.1–8.3%) and intra-laboratory reproducibility (CVRw = 4. 7–14.9%).	[30]
Irgacure-184 and Irgacure-651	Food simulant: 10% ethanol (v/v), 95% ethanol (v/v) and isooctane)	FID	LLE for 10% ethanol (v/v), 95% ethanol (v/v) Isooctane: concentration	Column: HP-5 (30 m × 0.32 mm × 0.25 μm) Injection mode: Splitless Oven temperature: 100 °C, then increased by 5 °C/min to up to 200 °C (2 min), then increased by 15 °C/min to up to 300 °C (5 min)	–	Low-density polyethylene (LDPE) or polypropylene (PP) coating can delay migration at certain condition. However the barrier property of PP was better than that of LDPE.	[31]

Irgacure 184, 2-benzyl-2-(dimethylamino)-1-(4-morpholino phenyl)-1-butanone (Irgacure 369), 2-dimethylamino-2-(4-methyl-benzyl)-1-(4-morpholin-4- yl-phenyl)-1-butanone (Irgacure 379), 2-methyl-1-[4-(methylthio) phenyl]-2-(4-morpholinyl)-1-propanone (Irgacure 907), 2,4,6-trimethyl benzoyldiphenyl phosphine oxide (TPO), 4-phenylben zophenone (PBZ), Speedcure 7010, Speedcure 7005, BP, ethyl-4-(dimethyl amino) Benzoate (EDMAB), ITX Omnipol 910	FCM	DART/TOF-MS time-of-flight mass spectrometry	LLE	Column: HP 5 MS (30 m × 0.25 mm × 0.25 µm) Injection mode: splitless Oven temperature: 3 min at 50 °C, followed by a 10 °C /min to 295 °C (2.5 min)	Above or below the 1.7 µg/dm^2/ (–)	Speed-cure 7005 and 4MBP were the compounds most susceptible to set-off in each of the samples by DART response.	[32]

(continued)

Table 5.3 (continued)

GC methods							
Analytes	**Matrix**	**Detector**	**Sample preparation or extraction**	**Column/injection mode/oven temperature**	**LOD/(LOQ)**	**Results**	**References**
BP, 2-methyl benzophenone (2-MBP), Irgacure 184, 3-methyl benzophenone (3-MBP), 4-MBP, 2-ethylhexyl-4-dimethylamino benzoate (EDB), Irgacure 651, o-methyl-benzoyl benzoate (OMBB), 2-ethylhexyl-4-dimethylamino benzoate (EHA) and ITX.	FCM: plastic materials PP, polyethylene terephthalate/cast polypropylene (PET/CPP), biaxially oriented polypropylene/cast polypropylene (BOPP/CPP), biaxially oriented polypropylene/low density polyethylene (BOPP/LDPE), and biaxially oriented polypropylene/polyethylene terephthalate/ cast polypropylene (BOPP/PET/CPP)	MS	Ultrasonic extraction followed by gel permeation chromatography (GPC)	Column: (30 m × 0.25 mm × 0.25 μm) Injection mode: split Oven temperature: 60 °C (1 min) to 200 °C at 40 °C/ min, then increased at 4 °C/ min to 280 °C (held for 3.5 min)	0.03×10^{-3} -0.12×10^{-3} mg/ dm^2/(–)	2-MBP, 3-MBP, Irgacure 184, and EHA were present in the samples. The highest level was of EHA in the BOPP/CPP (2.64×10^{-2} mg/ dm^2)	[33]

BP, EDMAB, 1-hydroxycyclohexyl phenylketone (HCPK), 4-MBP, 2-ethylhexyl-4-dimethylamino benzoate (EHDAB), 4-chlorobenzo phenone (p-CBP), 2-chlorothioxanthone (CTX), ITX; 2, 2-dimethoxy-2-phenyl acetophenone (2,2-DMPA), and OMBB.	Food: 12 brands of milk.	MS	SPME	Column: HP-5MS (30 m × 0.25 mm × 0.25 μm) Injection Mode: Splitless Oven temperature: 80 to 170 °C with an increasing rate at 25 °C/min (1 min), then increased to 220 °C with a rate of 5 °C/min (1 min), increased to 270 °C at a rate of 7 °C/min (5 min)	0.05–4.4 μg/L/ (–)	Three cases of positive samples were found. BP was detected in all three samples. EHDAB was detected in one sample, and ITX was detected in two samples.	[34]
BP, 2-MBP, Irgacure 184, EDMAB, N-ethyl-p-toluene sulfonamide, 2-hydroxy benzophenone (2-HBP), 3-MBP, 4-MBP, DMPA, methyl-2-benzoylbenzoate (MBB), 4-hydroxy benzophone, 2-ethylhexyl-4-(dimethylamino) benzophenone, Irgacure 907, 4-ITX, 2-ITX, PBZ, 2,2-diethyl-9H-thioxanthen-9-one (DETX), 4,4-bis (diethylamino) benzophenone (DEAB)	FCM: 11 paper or cardboard materials Food simulant: tenax	MS	Tenax extraction whit diethyl ether	Column: HP5MS column (30 m × 250 μm × 0.25 μm) Injection mode: Splitless Oven Temperature:100 °C (1 min), then the oven was programmed to rise at 10 °C/min to 300 °C (5 min)	0.007–0.176 mg/L /(0.024–0.179 mg/L)	Maximum values by specific migration test (at 40 °C for 10 days, using tenax) were 10.83 mg/L for BP, 0.53 mg/L for Irgacure 184, 0.47 mg/L for 4-MBP, 0.22 mg/L for DMPA and 0.4 mg/L for MBB.	[35]

(continued)

Table 5.3 (continued)

LC methods							
Analytes	**Matrix**	**Detector**	**Sample preparation or extraction**	**Column/mobile phase**	**LOD/(LOQ)**	**Results**	**Reference**
2-ITX	Food: 137 samples of fatty and non-fatty food FCM: plastic, carton, various	DAD/FLD	QuEChERS: Non-fatty foods FCM: extraction with hexafluoro-2-propanol and ethanol	Column: LC-PAH Supelcosil (250 × 4.6 mm × 5 μm) Isocratic mode Mobile phase: A: H2O; B: acetonitrile (ACN) H2O /ACN 15:85 v/v)	2 mg/μL/ (5 mg/μL)	2-ITX was detected in 36 of 137 packages. The levels of 2-ITX ranged up to 357 μg/kg in orange juice and 208 μg/kg in baby food.	[15]
2-isopropylthioxanthone (2-ITX) 4-isopropylthioxanthone (4-ITX)	Food: 37 milk samples, packaged in tetrapack	MS/MS	LLE	Column: Luna C8 (2) (50 × 2 mm × 5 μm) Gradient mode Mobile Phase: A: H_2O (0.05% acetic acid) B: ACN 40% to 80% of solvent B in 12 min, then to 100% of B in 2 min (4 min hold) Discovery ZR-PS (150 × 2.1 mm × 3 μm) Gradient mode	Luna C8 –/(2.5 ng/mL) Discovery ZR-PS –/(6.1–7.2 ng/m L)	16 samples were positive with concentrations ranging from 173 to 439 mg/L for 2-ITX and from < 6 to 25 mg/L for 4-ITX.	[36]

				Mobile phase: A: H2O (0.05% acetic acid) B: ACN 20% to 42% of solvent B in 36 min, then to 100% of B in 2 min (4 min hold)			
ITX	Food: 50 samples of yogurt, milk and pudding	MS	LLE	Column: Gemini C18 column (100 mm × 2.0 mm × 5 μm) Gradient mode Mobile phase: A (aqueous 20 mM ammonium formiate pH 4.5) B (methanol) The gradient program was: 0–3 min 70% B, from 3 to 5 min the B percentage moved from 70% to 90%, from 5 to 12 min B percentage was maintained at 90% B, from 12 to 13 min B percentage reached 99% B and was maintained at 99% from 13 to 15 min; from 15 to 16 min B percentage was brought back to the initial conditions 70%, and eventually the column was reconditioned at 70% B from 16 to 20 min	6.2 μg/kg / (7.2 μg/kg)	ITX was detected in 10 of 28 yoghurts samples: three samples contained a relevant concentration (330, 499 and 512 μg/kg).	[37]

(continued)

Table 5.3 (continued)

LC methods							
Analytes	**Matrix**	**Detector**	**Sample preparation or extraction**	**Column/mobile phase**	**LOD/(LOQ)**	**Results**	**Reference**
Irgacure 184,BP, Irgacure 651, Irgacure 907, ITX, EHDAB	Food: milk FCM: LDPE	DAD	LLE	Column: Kromasil 100 C18 (150 × 4 mm, 5 µm) Gradient mode Mobile phase A:Agua B: ACN 20% B–80% A (2 min hold), 20–100% of B in 21 minutes(7 min hold)	17–33 ng/mL/(–)	The diffusion coefficients at 5 °C ranged between 8.4×10^{-12} (for ITX) and 5.1×10^{-10} (for benzophenone) and those at 40 °C between 5.9×10^{-10} (for ITX) and 6.1×10^{-9} (for Irgacure 184).	[38]

BP	Food: 30 commercial packaged cakes FCM: PP, PP and ethylene vinyl alcohol (PP/EVOH/PP), polyethylene therephtalate, silic oxide and polyethylene (PET/SiOx/PE)	DAD GC-MS (confirmation)	LLE	Column: Kromasil 100 C18 (150 × 4 mm × 5 µm) Mobile Phase: A: H_2O B: ACN Gradient mode Mobil phase: ACN/water 50/50 to 100% ACN	FCM: 0.017 mg/L/(0.05 mg/L) Food: 0.15/(0.45 mg/kg)	PP plastics sleeves were found not to be an effective barrier against BP (3,800 µg/g of BP was found in cakes). In cakes wrapped with a multilayer film of PP/EVOH/PP (1,400 µg/g) and the concentration of BP in cakes wrapped with a multilayer film of PET/SiOx/PE was negligible.	[39]
2-Hydroxybenzo phenone (2-HBP), 4-hydroxybenzo phenone (4-HBP), 4-MBP, MBB, DEAB, PBZ, BP	Food: cake, bread, cereals, rice and pasta FCM	DAD	LLE	Column: Kromasil 100 C18 (250 × 4 mm × 5 µm) Gradient mode Mobile Phase: A: H_2O B: ACN 50%A/50%B (0 min); at min 10 min, 100% B; held at 100% B until min 15	–	Highest level of migration was found in the cake. Porosity and the fat content of the food have a strong influence on the migration process.	[40]

(continued)

Table 5.3 (continued)

LC methods							
Analytes	**Matrix**	**Detector**	**Sample preparation or extraction**	**Column/mobile phase**	**LOD/(LOQ)**	**Results**	**Reference**
BP, Irgacure 184, ITX, EHDAB, Irgacure 907, TPO, Irgacure 369	Food: milk	ESI*/MS/MS	SPE	Column: Gemini C18 (2.0 mm, 150 mm × 3 μm) Gradient mode Mobile phase: A: methanol B: mixture of 5 mmol/ L ammonium acetate and 0.1% formic acid. 0–3.5 min, 20–90% A; 3.5–25.0 min, 90–90% A; 25.0–26.0 min, 90–20% A; 26.0–29.0 min, 20–20% A	0.05–2.5 μg/kg / (0.1–5.0 μg/kg)	The most important contaminations were BP and ITX in concentration ranges of 2. 84–18.35 and 0.83–8.87 ug/kg, respectively.	[41]

ITX, 2-methyl-4′-(methylthio)-2-morpholinopropiophenone (MTMP), EDMAB	Food simulants: 50% ethanol (v/v) Food: yoghurt	DAD/FL	Yoghurt extracted with acetonitrile followed by following a liquid–liquid clean-up Etranol 50%	Column: Luna C18(2) (100 × 2 mm × 3 mm) Gradient mode Mobile phase: demineralized water/ACN: 0 min 40% ACN, 12 min 85% ACN, 16 min 100% ACN, 18 min 100% ACN	Migration into yogurt: ITX: < 0.4 μg/ kg EDMAB: < 3.5 μg/ kg/ (10.7 μg/ kg) MTMP: < 3.8 μg/ kg Migration into ethanol 50%: ITX: 1.4 μg/ kg/(5.8 μg/ kg) EDMAB: 2.5 μg/ kg/(8.2 μg/ kg) MTMP: 1.6 μg/ kg/(5.4 μg/ kg)	Migration of ITX and EDMAB increases with lamp age and storage time of the unfilled cups. 50% ethanol is a suitable simulant for yoghurt.	[14]

(continued)

Table 5.3 (continued)

LC methods							
Analytes	**Matrix**	**Detector**	**Sample preparation or extraction**	**Column/mobile phase**	**LOD/(LOQ)**	**Results**	**Reference**
BP, 4-MBP, PBZ, methyl-o-benzoylbenzoate (MOBB), 4-HBP, 4-dimethylaminobenzophenone (DMABP), 4,4′-bis (Dimethylamino)-benzophenone (Michler's ketone) (MK), 4,4′-bis(Diethylamino)-benzophenone (Michler's ethylketone) (MEK), (EDAB), EHDAB, HCPK	Food: 310 products predominately packed in cartonboard FCM	DAD (FCM) MS (Food) MS/MS (Food)	LLE	Column: Star RP-18e (250 mm × 4 mm × 5 µm) Gradient mode Mobile phase: water and ACN (A): 12 min isocratic at 30% A, linear gradient to 40% A in 1 min, 14 min isocratic at 40% A, linear gradient to 50% A in 5 min, 23 min isocratic at 50% A, linear gradient to 100% A in 5 min, 10 min isocratic at 100% A (stop time), linear gradient to 30% A in 0.5 min, 7 min isocratic at 30% A	HPLC-DAD 2. 8–29 µg/dm^2/(9. 4–95 µg/dm^2) HPLC-MS 4.8–195 µg/kg / (17–625 µg/kg) HPLC-MS/MS 2.5–38 µg/kg / (7.5–113 µg/kg)	BP was detected in 49% of the packaging materials, followed by MBP (8%), HCPK (7%) and MOBB (5%). BP, HCPK, MBP and MOBB were found in 20 foodstuffs (one or more of these compounds) in quantities above the legally acceptable limits.	[42]

BP, MBB, 2-HBP, 4-MBP, PBZ, EHA, EDMAB, ITX, Irgacure 184, Irgacure 651, 2-hydroxy-2-methylpropiophenone (Irgacure 1173), Benzophenone acrylate (BPAcr), DEAB, DETX	FCM: 17 samples (can, paper, cardboard, and plastic)	DAD MS/MS	LLE	HPLC:DAD Column: Kromasil C18 (250 mm × 3.2 mm × 5 μm) Gradient mode Mobile phase: A: water. B: ACN The first 2 min 40:60 (%, v/v) A–B, increased to 100% B in 23 min and then maintained constant until 30 min HPLC-MS/MS Column: Kromasil C18 (150 mm × 3.2 mm × 5 μm) Gradient mode Mobile Phase: A: ACN containing 0.1 % (v/v) formic acid B: Milli-Q water containing 0.1%(v/v) formic acid	0.31–15.6 μg/ dm^2/(0.78–31.2 μg/dm^2)	Photinitiators were detected in 47% of the samples tested (ranged 2.23 to 102.93 μg /dm^2) MBB, 2-HBP, 4-MBP, ITX, Irgacure 651, Irgacure 1173 and BPAcr were not found	[43]

(continued)

Table 5.3 (continued)

LC methods							
Analytes	**Matrix**	**Detector**	**Sample preparation or extraction**	**Column/mobile phase**	**LOD/(LOQ)**	**Results**	**Reference**
BP, DEAB, CTX, 1-chloro-4-ropoxy-9H-thioxanthen-9-one (CPTX), DMPA, 4-(dimethylamino) benzophenone (DMBP), 2-ethylanthraquinone (EA), EHDAB, EDMAB, 4-HBP, HCPK, 2-hydroxy-4-methoxybenzo phenone (HMBP), 2-hydroxy-4′-(2-hydroxyethoxy)-2-methylpropiophenone (HMMP), ITX, 4-MBP, MK, PBZ	Food: 97 dry foodstuffs	MS/MS	SPE	Column: UPLC BEH C18 (100 mm × 2.1 mm × 1.7 μm) Gradient mode Mobile phase: A: water with 0.1% formic acid B: ACN with 0.1% formic acid 40% B to 100% B in 6 min. This was kept constant for 2 min (column was conditioned for 2 min with the initial mobile phase composition)	0.01–25 μg/kg / (0.1–75 μg/kg)	At least one photoinitiator was detected in 89% of the samples. BP, EDMAB and DMPA were most frequently found, while CPTX, DEAB, DMBP, HCPK, HMBP and MK were never detected.	[44]

HMMP, 4-HBP, EDMAB, BP, DMBP, MK, HMBP, 4-MBP, EA, DEAB, CTX, CPTX, ITX, EHDAB	Food simulant Food: cereal and rice samples	MS/MS	Extraction LLE	Column: UPLC BEH C18 (100 mm × 2.1 mm × 1.7 µm) Gradient mode Mobile phase: A: water with 0.1% formic acid B: ACN with 0.1% formic acid gradient from 40% B to 100% B in 6 min. This was kept constant for 2 min	–	The simulation with Tenax at 60 °C overestimated migration in cereals up to a maximum of 92%. Tenax is a much stronger adsorbent than rice and cereals.	[45]
BP, MK, 4-MBP, PBZ, DEAB	Food simulants: Tenax and 95% ethanol (v/v)	DAD	Tenax: extracted with ethanol. 95% ethanol: direct injection	Column: Venusil MP C18 (250 × 4.6 mm) Gradient mode Mobile Phase: A: water B: ACN 0–12 min, 60–75% B; 12–13 min, 75–80% B; 13–17 min, 80% B; 17–18 min, 80–60% B.	–	The amount of photoinitiators that migrated into the Tenax increased with the contact time at each temperature. Equilibrium times in 95% ethanol (v/v) were much shorter than those achieved in Tenax.	[46]

* Electrospray ionization
FCM: Food contact materials
(–): Not available

EHDAB] in milk samples. The extract was analysed by HPLC–UV and results were confirmed using LC–MS.

Conventional SPE was also reported for photoinitiator determination; SPE was applied to extract five ink photoinitiators (ITX, BP, EHDAB, Irgacure® 184 and EDMAB) in packaged food beverages. GC–MS was used for analyte quantification and LC–MS/MS to confirm the presence of BP [51].

SPME is a solvent-free extraction technique chosen by some authors. Negreira et al. [52] developed a method based on the SPME technique for the determination of seven ink photoinitiators in packed milk samples. A GC–MS system was used for analyte determination. Using microextraction techniques, performance was affected by the characteristics of the sample; however, no differences were observed among different milks with the same nominal fat content. Recently, Liu et al. [34] used SPME combined with a simple method of protein precipitation prior to GC–MS analysis to simultaneously determine 10 photoinitiators in milk.

Other extraction techniques such as PLE have also been used for the analysis of these compounds. Sagratini et al. [53] developed a method for ITX determination in fruit juices, based on the use of PLE, prior to LC–ESI–MS analysis. PLE conditions were optimised and recovery values of 73% were obtained using a mixture of hexane/acetone (50:50) with a 50% flush volume at 100 °C and 1,500 psi for 5 min static time in one cycle.

Gil-Vergara et al. [54] checked the extraction efficiency of PLE and conventional LLE for the analysis of ITX and EHDAB in milk and milk-based beverages. Results showed that PLE using ethyl acetate as a solvent without a further purification method was suitable for milk samples with good recoveries.

Nowadays, because of its simplicity and a minimal number of steps, some authors have used the QuEChERs method to determine photoinitiators in food matrices. Gallart Ayala et al. [55] employed an LC–MS/MS method using a QuEChERS extraction for the determination of 11 UV photoinitiators in packaged foods.

Other extraction techniques such as dispersive liquid–liquid microextraction (DLLME) and cloud point extraction (CPE) have also been used for the analysis of photoinitiators.

DLLME is a simple novel method, based on the dispersion of an extraction solvent and a disperser solvent in an aqueous solution. Zhang and Jiao [49] developed a technique composed of DLLME and MEKC with DAD for the determination of seven photoinitiators in fruit juice. Methanol was chosen as the dispersant and 1,1,2-trichloroethane ($C_2H_3Cl_3$) as the extractant. Recoveries from 85.6% to 124.7% were obtained.

CPE is another extraction alternative for photoinitiators, which is a promising technique that employs nonionic surfactants, with advantages such as a high preconcentration factor and relatively low toxicity. Thus, as an alternative to SPE, QuEChERS and LLE methods, Ding et al. [56] developed CPE based on Tween® 20 for the extraction and preconcentration of 4-MBP in milk samples prior to analysis by HPLC. The recovery of 4-MBP was satisfactory, and no matrix interference was observed in the complex milk samples.

In addition to the analysis of photoinitiators in foodstuffs, the migration of these compounds in food simulants from plastic and paper materials has also been reported. Because of the complexity of food matrices, migration assays are performed using food simulants, which are less complex than food. However, migration studies in both foodstuffs and food simulants are expensive and resource consuming. In order to avoid these procedures, it is possible to use mathematical diffusion models to predict the migration process. One of the most commonly used mathematical models for studying the migration process in food-packaging materials is derived from Fick's second law, which describes the diffusion phenomena. Mathematical models allow the calculation of two key parameters in the migration process: the diffusion coefficient (the rate at which the migrant moves in the polymer matrix) and the partition coefficient between the polymer and the food (the thermodynamic equilibrium of the migration process) [57, 58].

Barnkob and Petersen [59] studied the migration of BP from paperboard into the food simulant Tenax. Kinetic migration was performed at a constant temperature of 34 °C with three relative humidities between 39% and > 73%. Quantification of BP was performed using GC–MS. Diffusion and partition coefficients were calculated using MIGRATEST software for modelling migration in multilayer materials. Results showed that BP migration after more than 30 days was higher at a relative humidity (RH) of > 73% compared with a RH of 64–71%. In the same way, diffusion and partition coefficients between paperboard and Tenax decrease with increasing RH.

Zhang et al. [60] developed a method based on supercritical fluid chromatography combined with a photodiode array detector and MS/MS to evaluate the migration of 13 photoinitiators [BP, ahydroxycyclohexylphenylketone (Darocure® 184), EDMAB, 4-MBP, 2,2-dimethoxy-2-phenylacetophenone (Irgacure® 651), OMBB, 2-ITX, PBZ, 2,2-diethoxyacetophenone, phenylbis(2,4,6-trimethylbenzoyl)phosphine oxide (Irgacure® 819), 2-hydroxy-2-methyl propiophenone (Darocure® 1173), TPO and Irgacure® 907] from PE food packaging using five different food simulants (10% ethanol, 3% acetic acid, 20% ethanol, 50% ethanol and vegetable oil). Results showed that among all 13 photoinitiators, Irgacure® 907 had the higher migration rates, while Darocure® 1173, Irgacure® 819 and TPO had the lowest migration rates. Migration kinetics of photoinitiators showed that migration rates increased with increasing contact time between the packaging and food simulant. In the same way, the character of the migrants also influenced the migration process.

In order to better understand the mass transfer process, Maia et al. [57] studied the migration of BP from LDPE to determine the diffusion and partition coefficients. Migration assays were performed at different time–temperature conditions in real foodstuffs: liquid food (orange juice, red wine and tomato sauce), solid food (Gouda cheese, turkey ham and cooked ham) and semisolid food with high viscosity (chocolate spread and pâté). A HPLC–DAD method was used for the analysis of

BP. The migration process was simulated using a mathematical model based on Fick's second law of diffusion. Results showed that the physical state and temperature of food affected the diffusion coefficients, while the partition coefficient was affected by, among other things, the food fat content.

In 2017, Cai et al. [61] evaluated the migration of four photoinitiators (BP, EHDAB, 4-MBP and Irgacure® 907) from four food-packaging materials (kraft paper, white cardboard, PE-coated paper and composite paper) using Tenax as the food simulant. To measure the photoinitiator content in Tenax, an LC–MS/MS method was developed. The Weibull model was used to predict the migration of photoinitiators in paper-packaging materials. BP was the compound that showed the highest migration rates, which was to be expected since BP was the smallest of the four compounds. Results showed that the migration process is influenced by the lipophilicity and the MW of photoinitiators.

UV printing inks are widely used in the printing of plastics and metal sheets (nonabsorbent materials) and are mainly composed of monomers, oligomers, pigments, additives and photoinitiators. Because of the different requirements of the finished printed product, the content of the individual ink component varies considerably; thus, it is also important to consider toxicological and migration properties.

The risk of the packed food becoming contaminated by one or several of the printing ink components is always present. Add to this the fact that many of the substances used in inks for food packaging have not been evaluated with regard to their toxicological properties. Of the ink components, photoinitiators have perhaps been the substances most evaluated; thus, they have been found in a variety of food products as a result of their migration from printed packaging materials.

The food-packaging industry, including inks, is continuously innovating, so that new materials used in the formulation will need to be assessed for toxicological and migration potential.

References

[1] M.J. Forrest in *Coatings and Inks for Food Contact Materials*, Rapra Review Report No.186, Smithers Rapra, Shawbury, UK, 2005.

[2] S. Chatani, C.J. Kloxin and C.N. Bowman, *Polymer Chemistry*, 2014, 5, 7, 2187.

[3] J.P. Fouassier, X. Allonas and D. Burget, *Progress in Organic Coatings*, 2003, 47, 1, 16.

[4] M.A. Lago, A Rodriguez-Bernaldo. de Quiros, R. Sendón, J. Bustos, M.T. Nieto and P. Paseiro, *Food Additives & Contaminants: Part A*, 2015, 32, 5, 779.

[5] T.M. Lovestead, A.K. O'Brien and C.N. Bowman, *Journal of Photochemistry and Photobiology A: Chemistry*, 2003, 159, 2, 135.

[6] W.A. Green in *Industrial Photoinitiators: A Technical Guide*, Taylor & Francis, Boca Ratón, FL, USA, 2010.

[7] M. Szycher in *Szycher's Handbook of Polyurethanes*, 2nd Edition, Taylor & Francis, Boca Raton, FL, USA, 2012.

[8] J. Ortyl and R. Popielarz, *Polimery*, 2012, 57, 7–8, 510.

[9] European Commission Regulation (EU) No 10/2011 of 14th January 2011 on Plastic Materials and Articles Intended to Come into Contact with Food, *Official Journal of the European Union L*, 2011, 12, 1.

[10] *Swiss Ordinance on Materials and Articles in Contact with Food*, (817.023.21), Swiss Federal Department of Home Affairs, Bern, Switzerland, May 2017. http://eupia.org/uploads/tx_edm/2017-10-05_Swiss_FCM_Ordinance_-_EuPIA_Q_A_regarding_non-DFC_FCM_Inks.pdf

[11] *EuPIA Suitability List of Photo-initiators for Low Migration UV Printing Inks and Varnishes*, European Printing Ink Agency, Brussels, Belgium, February 2013. http://www.eupia.org/uploads/tx_edm/130219_corr3_EuPIA_Suitability_List_of_Photoinitiators_for_Low_migration_UV_Printing_Inks_and_Varnishes.pdf

[12] T. Richter, T. Gude and T. Simat, *Food Additives & Contaminants*, 2009, 26, 12, 1574.

[13] M.A. Lago, R. Sendón, A. Rodríguez-Bernaldo de Quirós in *Photocured Materials*, Eds., A. Tiwari and A. Polykarpov, The Royal Society of Chemistry, Cambridge, UK, 2015, p.290.

[14] T. Jung, T.J. Simat and W. Altkofer, *Food Additives & Contaminants: Part A*, 2010, 27, 7, 1040.

[15] T. Rothenbacher, M. Baumann and D. Fügel, *Food Additives & Contaminants*, 2007, 24, 4, 438.

[16] T. Jamnicki and S. Jamnicki, *Acta Graphica Znanstveni Časopis Za Tiskarstvo I Grafičke Komunikacije*, 2011, 21, 1–2, 7.

[17] J.L. Aparicio and M. Elizalde, *Packaging Technology and Science*, 2015, 28, 3, 181.

[18] *RASFF – Food and Feed Safety Alerts*, Rapid Alert System for Food and Feed. https://ec.europa.eu/food/safety/rasff_en

[19] *Toxicological Evaluation of Benzophenone* in *EFSA Journal*, 2009, 1104, 1.

[20] *2-Isopropyl Thioxanthone (ITX) and 2-Ethylhexyl-4-Dimethylaminobenzoate (EHDAB) in Food Contact Materials*, Question No: EFSA-Q-2005-240 & EFSA-Q-2005-241, *EFSA Journal*, 2005, 293, 1. http://www.efsa.europa.eu/sites/default/files/scientific_output/files/main_documents/293.pdf

[21] M. Reitsma, T.F. Bovee, A.A. Peijnenburg, P.J. Hendriksen, R.L. Hoogenboom and J.C. Rijk, *Toxicological Sciences*, 2012, 132, 1, 64.

[22] T. Jung and T.J. Simat, *Food Additives & Contaminants: Part A*, 2014, 31, 4, 743.

[23] C. Sun, S.H. Chan, D. Lu, H.M.W. Lee and B.C. Bloodworth, *Journal of Chromatography A*, 2007, 1143, 1, 162.

[24] R. Koivikko, S. Pastorelli, A. Rodríguez-Bernaldo de Quirós, R. Paseiro-Cerrato, P. Paseiro-Losada and C. Simoneau, *Food Additives & Contaminants*, 2010, 27, 10, 1478.

[25] E.L. Bradley, J.S. Stratton, J. Leak, L. Lister and L. Castle, *Food Additives & Contaminants: Part B*, 2013, 6, 2, 73.

[26] W.A.C. Anderson and L. Castle, *Food Additives & Contaminants*, 2003, 20, 6, 607.

[27] G. Allegrone, I. Tamaro, S. Spinardi and G. Grosa, *Journal of Chromatography A*, 2008, 1214, 1, 128.

[28] Z.W. Wang, X.L. Huang and C.Y. Hu, *Packaging Technology and Science*, 2009, 22, 3, 151.

[29] S.M. Johns, S.M. Jickells, W.A. Read and L. Castle, *Packaging Technology and Science*, 2000, 13, 3, 99.

[30] E. Van Hoeck, T. De Schaetzen, C. Pacquet, F. Bolle, L. Boxus and J. Van Loco, *Analytica Chimica Acta*, 2010, 663, 1, 55.

[31] X.L. Huang, Z.W. Wang, C.Y. Hu, Y. Zhu and J. Wang, *Packaging Technology and Science*, 2012, 26, S1, 23.

[32] K. Bentayeb, L.K. Ackerman, T. Lord and T.H. Begley, *Food Additives & Contaminants: Part A*, 2013, 30, 4, 750.

[33] H. Cai, G. Shang, J. Zhang, G. Tao, G. Chen and X. Wan, *Analytical Methods*, 2015, 7, 21, 9026.

[34] P. Liu, C. Zhao, Y. Zhang and Y. Chen, *Journal of Food Science*, 2016, 81, 5, T1336.

[35] U. Altuntaş, V. Hitay and B. Özçelik, *Packaging Technology and Science*, 2016, 29, 10, 513.
[36] R. Bagnati, G. Bianchi, E. Marangon, E., Zuccato, R. Fanelli and E. Davoli, *Rapid Communications in Mass Spectrometry*, 2007, 21, 13, 1998.
[37] C. Benetti, R. Angeletti, G. Binato, A. Biancardi and G. Biamotto, *Analytica Chimica Acta*, 2008, 617, 1, 132.
[38] A. Sanches-Silva, S. Pastorelli, J.M. Cruz, C. Simoneau, I. Castanheira and P. Paseiro-Losada, *Journal of Agricultural and Food Chemistry*, 2008, 56, 8, 2722.
[39] S. Pastorelli, A. Sanches-Silva, J.M. Cruz, C. Simoneau and P.P. Losada, *European Food Research and Technology*, 2008, 227, 6, 1585.
[40] A. Rodríguez-Bernaldo de Quirós, R. Paseiro-Cerrato, S. Pastorelli, R. Koivikko, C. Simoneau and P. Paseiro-Losada, *Journal of Agricultural and Food Chemistry*, 2009, 57, 21, 10211.
[41] D.X. Shen, H.Z. Lian, T. Ding, J.Z. Xu and C.Y. Shen, *Analytical and Bioanalytical Chemistry*, 2009, 395, 7, 2359.
[42] T. Jung, T.J. Simat, W. Altkofer and D. Fügel, *Food Additives & Contaminants: Part A*, 2013, 30, 11, 1993.
[43] M.A. Lago, A.R.B de Quirós, R. Sendón, J. Bustos, M.I. Santillana and P. Paseiro, *Analytical and Bioanalytical Chemistry*, 2014, 406, 17, 4251.
[44] K. van den Houwe, A. van Heyst, C. Evrard, J. van Loco, F. Bolle, F. Lynen and E. Van Hoeck, *Packaging Technology and Science*, 2016, 29, 2, 121.
[45] K. van den Houwe, C. Evrard, J. van Loco, F. Lynen and E. van Hoeck, *Food Additives & Contaminants: Part A*, 2016, 33, 5, 913.
[46] B. Han, L. Ding, R. Su, L. Chen, L. Wang, W. Qi and Z. He, *Food Additives & Contaminants: Part A*, 2016, 33, 5, 876.
[47] G. Morlock and W. Schwack, *Analytical and Bioanalytical Chemistry*, 2006, 385, 3, 586.
[48] L.K. Ackerman, G.O. Noonan and T.H. Begley, *Food Additives & Contaminants*, 2009, 26, 12, 1611.
[49] Y. Zhang and B. Jiao, *Food Analytical Methods*, 2015, 8, 4, 973.
[50] A. Sanches-Silva, S. Pastorelli, J.M. Cruz, C. Simoneau, I. Castanheira and P. Paseiro-Losada, *Journal of Dairy Science*, 2008, 91, 3, 900.
[51] G. Sagratini, G. Caprioli, G. Cristalli, D. Giardiná, M. Ricciutelli, R. Volpini, Z. Yanting and S. Vittori, *Journal of Chromatography A*, 2008, 1194, 2, 213.
[52] N. Negreira, I. Rodríguez, E. Rubí and R. Cela, *Talanta*, 2010, 82, 1, 296.
[53] G. Sagratini, J. Manes, D. Giardiná and Y. Picó, *Journal of Agricultural and Food Chemistry*, 2006, 54, 20, 7947.
[54] A. Gil-Vergara, C. Blasco and Y. Picó, *Analytical and Bioanalytical Chemistry*, 2007, 389, 2, 605.
[55] H. Gallart-Ayala, O. Núñez, E. Moyano and M.T. Galceran, *Journal of Chromatography A*, 2011, 1218, 3, 459.
[56] S. Ding, Y. Cao, A. Gong and Y. Wang, *Journal of Separation Science*, 2016, 39, 20, 4027.
[57] J. Maia, A. Rodríguez-Bernaldo de Quirós, R. Sendón, J.M. Cruz, A. Seiler, R. Franz, C. Simoneau, L. Castle, M. Driffield, P. Mercea, P. Oldring, V. Tosa and P. Paseiro, *Food Additives & Contaminants: Part A*, 2016, 33, 4, 715.
[58] *Applicability of Generally Recognized Diffusion Models for the Estimation of Specific Migration in Support of EU* Directive *2002/72/EC*, Ed., C. Simoneau, JRC Publications, 2010, EUR 24514 EN. http://publications.jrc.ec.europa.eu/repository/handle/JRC59476
[59] L.L. Barnkob and J.H. Petersen, *Food Additives & Contaminants: Part A*, 2013, 30, 2, 395.
[60] Y. Zhang, Z. Du, X. Xia, Q. Guo, H. Wu and W. Yu, *Polymer Testing*, 2016, 53, 276.
[61] H. Cai, S. Ji, J. Zhang, G. Tao, C. Peng, R. Hou, L. Zhang, Y. Sun and X. Wan, *Food Additives & Contaminants: Part A*, 2017, 34, 1632.

6 Other chemical substances from food contact materials

6.1 Other bisphenol-related compounds

Bisphenol A (BPA) is a chemical compound that has been the focus of concern over the last decade, regarding its use in food contact materials (FCM). Analogue compounds, that is, molecules containing two hydroxyphenyl groups, are suspected of being used instead of BPA, and therefore could be present in food if migration occurs. Some examples are shown in Figure 6.1.

Bisphenol F {4-[(4-hydroxyphenyl)methyl]phenol, CAS 620-92-8} is the lower molecular weight (MW) monomer present in phenol formaldehyde resins. Moreover, it is used in plastic adhesives and other nonfood contact applications [1].

Bisphenol S (BPS) (4,4′-sulphony diphenol, CAS 80-09-1) has two hydroxyphenyl groups linked by a sulfonyl group and has been suspected of being used to replace BPA. According to Regulation European Union (EU) 10/2011, BPS is authorised for use as a monomer to produce FCM plastics and has a specific migration limit of 0.05 mg/kg [2].

For all other bisphenol analogues, there is a lack of information regarding their use and industrial production; however, most of them have been found in several foods and due to the similar toxic effects to BPA, further research is needed concerning the presence of these compounds in foods and FCM [1].

For the determination of these bisphenolic compounds, high-performance liquid chromatography with ultraviolet (UV) and/or fluorescence detectors could be used, but liquid chromatography (LC)–tandem mass spectrometry MS/MS (using negative electrospray ionisation) has shown its applicability in items such as household waste paper [3], biological samples [4] and foodstuffs [5], and could be used to investigate their presence in FCM.

6.2 Fluorinated substances (perfluorochemicals)

Perfluorinated compounds, namely, poly- and perfluoroalkyl substances (PFAS), have hydrophobic and oleophobic properties and are used because they serve as oil and water repellents. Moreover, they possess great thermal and chemical stability. For this reason, they are employed in a wide variety of applications, comprising, for instance, coatings made by polytetrafluoroethylene (PTFE), which is a material with a high melting point (327 °C) and so confers stability and resistance [6]. From a FCM point of view, PFAS are mainly present in nonstick cookware and coated papers [7].

https://doi.org/10.1515/9783110648065-006

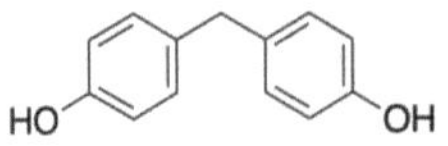

Bisphenol F (620-92-8)

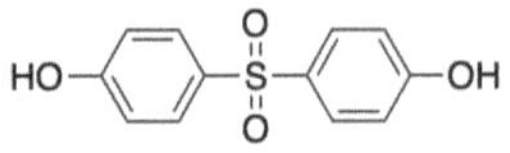

Bisphenol S (CAS 08-09-1)

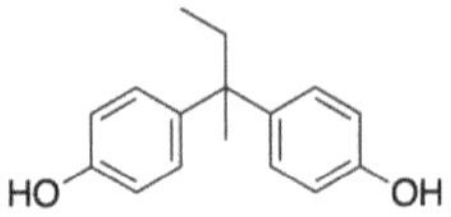

Bisphenol B (77-40-7)

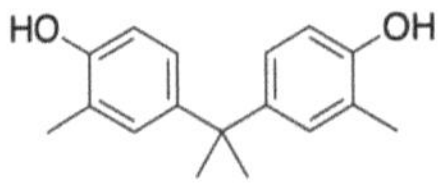

Bisphenol C (79-97-0)

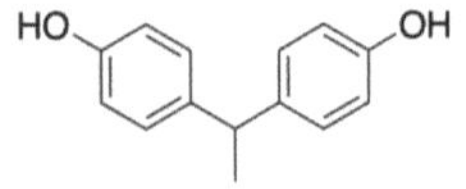

Bisphenol E (2081-08-5)

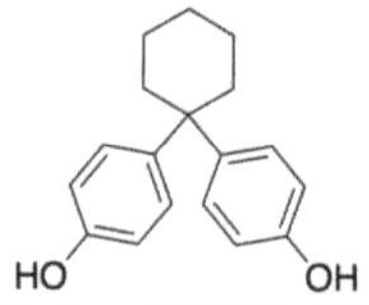

Bisphenol Z (843-55-0)

Figure 6.1: Some analogue compounds related to BPA.

PFAS are a very large group of different chemical compounds; perfluorooctane sulfonate (PFOS) and perfluorooctanoic acid (PFOA) are the most important compounds of PFAS and the perfluoroalkyl carboxylates (PFCA), respectively. They have been described as persistent compounds and accumulate in food and the environment, and are thus a health concern. In fact, PFOS production and use has been banned by the Stockholm Convention on organic pollutants since 2009 [8, 9]. Fluorinated telomer alcohols (PTOH) and polyfluoroalkyl phosphates (PAP) are used as precursors in the manufacture of PFAS and should also be taken into consideration when PFAS are monitored and/or evaluated [10]. PFAS are also used in fluorochemical-treated paper that is extensively used in the fast-food industry due to its improved resistance [11]. The bioaccumulation of long-chain PFAS (alkyl chain of eight carbons or longer) has been demonstrated in the liver, immune system and reproductive organs, where the most toxicity was observed. PFCA is a family of compounds with well-characterised toxic effects; however, PTOH needs further evaluation [9]. Disruption of reproductive development and the endocrine system as well as tumour promotion were some toxicological effects reported for PFOS congeners [10].

Low MW PFAS (short-chain compounds, i.e., less than C_6) could be an alternative because they are not bioaccumulative; however, their toxicity needs to be further explored [9].

Some works suggest that fluoropolymer FCM are not the main source of food contamination, whereas coated paper (e.g., popcorn bag paper and hamburger bags) used in contact with warm food could contribute to the presence of PFAS in food [6, 11].

PFOA was detected in microwave popcorn bag papers in higher quantities (0.3 mg/kg bag) than in PTFE cookware [6]; cookware treated with fluoropolymers has also been analysed [12], but PFOA was not detected after cooking experiments. Recently, different samples (wrapping paper, breakfast bags, baking paper and roasting bags) were analysed and PFCA and PFAS were detected in the range of 0.02–6.2 pg/cm^2, with PFOA being the most frequently identified PFCA and PFOS the most frequently identified PFAS [13].

PFAS do not have any chromophore group and are nonvolatile; therefore, LC–tandem mass spectrometry (MS/MS) is the analytical technique most utilised to quantify these compounds at very low levels in foods and also in FCM [8].

Zabaleta et al. recently identified and quantified numerous PAP, along with their intermediate and degradation products. Besides several PFAS, different chain length fluorotelomer saturated acids and fluorotelomer unsaturated acids, considered intermediates of PFAS precursors, were detected in food contact samples. Among the PFCA, those of short chains (C_4–C_8) were detected in samples collected in Europe and America (up to 820 ng/g), while samples from Asia contained mainly long-chain PFCA (C_8–C_{16}) (PFOA up to 56 ng/g). This reveals that while in Europe and America long-chain PFCA are being replaced by short-chain PFCA, in Asia the presence of long-chain PFCA is still an unresolved issue [14].

6.3 Primary aromatic amines

The migration of primary aromatic amines (PAA) has been reported from polyurethane (PU)-based laminates [15] and also from kitchen utensils [16]. The occurrence of PAA can originate from the use of diisocyanates in PU adhesives or in the azo dye colourants employed in Nylon kitchen utensils [6].

Some of these aromatic compounds are classified by the International Agency for Research on Cancer in group 1 (carcinogenic to humans), group 2A (probably carcinogenic to humans) and group 2B (possibly carcinogenic to humans) [17]. Figure 6.2 shows some examples of PAA.

PU adhesives, used in laminated films, printing inks and lacquers, are polymerisation products of polyols and diisocyanate monomers. PAA could be present in food if the free isocyanate aromatic monomer persists after the polymerisation process, migrating to the food and producing aromatic amines if water is present [7].

Migration assays have shown that PAA are also found in Nylon kitchen utensils. These utensils (frequently black) can contain azo dyes that are organic colourants containing chromophore azo groups that are prepared from aromatic amines [16]. 4,4′-MDA and aniline were the PAA found at higher concentrations in most migration experiments. It has also been suggested that 4,4′-MDA could be used during polyamide production to improve its thermal stability [18].

European legislation has established that the levels of PAA in food or food simulants should not be detectable (i.e., 0.01 mg/kg), except for those that are listed in Regulation EU 10/2011 [European Commission (EC), 2011] [2]. Since the first case was reported in 2004 to date, approximately 300 notifications have been published by different European countries reporting high levels of these compounds, mainly in utensils imported from Asia [19].

Name	Abbreviation	CAS N°	Structure	IARC Group
2,4-Toluenediamine	2,4-TDA	95-80-7	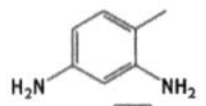	2B
o-Toluidin	o-T	62-53-3	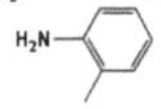	1
4,4′-Diaminodiphenylether	4,4′-DPE	101-80-4	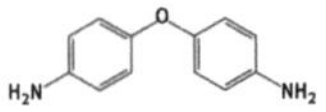	2B
Benzidine	BNZ	92-87-5	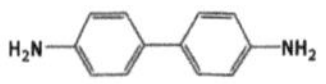	1
4,4′-Methylenedianiline	4,4′-MDA	101-77-9	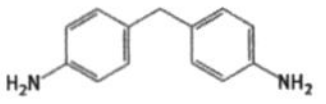	2B
3,3′-Dimethylbenzidine	3,3′-DMB	119-93-7	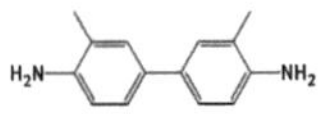	2B

Figure 6.2: Examples of PAA.

There are several published analytical methods that achieve the detection limit required by European legislation. A spectrophotometric method has been used, as well as chromatographic methods including LC–UV and gas chromatography (GC)–MS, with LC coupled to MS gaining success due to the unequivocal identification of the PAA detected and the low detection limits reached [7]. However, over the last few years, high-resolution MS in combination with LC has been used for the analysis of these compounds [8].

6.4 Mineral oil hydrocarbons

Mineral oil hydrocarbons (MOH) are a varied group of substances that include molecules containing saturated and/or unsaturated hydrocarbons and display a linear, branched or cyclic structure; they are very complex mixtures with different carbon atoms and structures. The term mineral oil saturated hydrocarbons (MOSH) involves open chain alkanes (straight and branched), that is, paraffins and naphthenes (largely alkylated cycloalkanes); mineral oil aromatic hydrocarbons (MOAH) are also included in MOH. Moreover, they are available from technical to food and analytical grades, with different compositions mainly regarding the MOAH content (technical-grade MOH contains 15–35% MOAH, whereas the MOAH content is negligible in food-grade MOH) [20].

A large number of products include MOH, for example, cleaning agents, personal care products and so on. They are used not only in FCM, mainly recycled paper and board, but also in plastics, adhesives and printing inks, among others. Thus, FCM is one of the sources of exposure to these compounds and, due to the above-mentioned uses, MOH are environmental contaminants. Furthermore, food-grade MOH are listed as food additives (E905) and are also used during the processing of foods (i.e., release agents) [20].

The migration of these compounds from FCM was recently related to paper and cardboard, often recycled, and the migration process involves the gas-phase transport of the compounds, with dry foods seemingly the most susceptible [21]. The German Federal Ministry of Food and Agriculture has elaborated a draft to establish the maximum content of MOH components in recycled cardboard for food contact applications: 2 mg/kg for MOSH (compounds between n-C_{20} and n-C_{35}) and 0.5 mg/kg for MOAH (compounds between C_{16} and C_{35}) [8].

In line with this, the EC published Recommendation EU 2017/84 [22] asking the Member States to monitor MOH in a wide variety of foods and also in their FCM. Thus, the European Food Safety Agency (EFSA) should be informed of the obtained results; moreover, all interested parties (i.e., food manufacturers, processors and so on) have been asked to provide the EFSA with all data available on MOH.

6.5 Nanoparticles

Nanoparticles (NP) (all external dimensions in the size range of 1–100 nm), when incorporated in a material, modify and frequently enhance material characteristics such as transparency, gas and moisture barrier properties, and antimicrobial properties. For this reason, they have been included in different products by several industrial sectors including medical, cosmetics, electronics and so on. Within the food-packaging sector, engineered nanoparticles (ENP) are often incorporated into active and intelligent packaging designed to increase the shelf life of foods and to improve its safety [23].

The addition of ENP to the polymer can be via embedded, fixed or bonded forms. Besides ENP, nanofibres (the third dimension is larger) or nanoplates (two external dimensions are larger) can also be incorporated [24].

The main question that arises from the use of ENP in FCM is if these particles are able to reach the food in contact with the FCM. If migration occurs, the risk of the use of these FCM needs to be evaluated [25].

Numerous publications refer to the evaluation of silver migration from polymer nanosilver composites, but the results reported are quite different. Some authors suggest that one reason for such variable migration values could be the special redox behaviour of silver (that could lead to the formation of other chemical species) and/or the physical alteration of the food contact surface (e.g., by an impact) [26].

Regarding the analysis of NP migration from FCM, the most commonly used analytical technique is inductively coupled plasma (ICP)–MS due to the low concentrations of NP expected; however, it has the drawback that it can only determine the total elemental concentration in the migration solution. By using single particle ICP–MS, the ionic and nanoforms of a particle can be discriminated and particle size distributions obtained. Moreover, no sample preparation is needed; however, this analytical technique has only been applied to migration solutions. Electron microscopy techniques have also been applied to evaluate the migration of NP; in these cases, information regarding the size and shape of the NP is obtained and furthermore, smaller particles can be detected. Because of the higher concentrations, some additional steps in sample preparation could be required (centrifugation and evaporation), which can create artefacts that lead to overestimated or underestimated results [24].

References

[1] D. Chen, K. Kannan, H. Tan, Z. Zheng, Y. Feng, Y. Wu and M. Widelka, *Environmental Science & Technology*, 2016, 50, 5438.
[2] European Commission Regulation (EU) No 10/2011 of 14th January 2011 on Plastic Materials and Articles Intended to Come into Contact with Food, *Official Journal of the European Union L*, 2011, 12, 1.
[3] G.A. Pivnenko, E. Pedersen, E. Eriksson and T.F. Astrup, *Waste Management*, 2015, 44, 39.
[4] B.A. Rocha, B.R. da Costa, N.C. de Albuquerque, A.R. de Oliveira, J.M. Souza, M. Al-Tameemi, A.D. Campiglia and F. Barbosa, Jr., *Talanta*, 2016, 154, 511.
[5] C. Liao and K. Kannan, *Journal of Agricultural and Food Chemistry*, 2013, 61, 4655.
[6] T.H. Begley, K. White, P. Honigfort, M.L. Twaroski, R. Neches and R.A. Walker, *Food Additives & Contaminants*, 2005, 22, 1023.
[7] R. Sendón and A. Sanches Silva in *Food Toxicology*, CRC Press, Boca Raton, FL, USA, 2016, p.461.
[8] Y. Sanchis, V. Yusà and C. Coscollà, *Journal of Chromatography A*, 2017, 1490, 22.
[9] P.A. Rice, O.J. Bandele and P. Honigfort in *Toxicants in Food Packaging and Household Plastics*, Ed., S. M. Snedeker, Springer-Verlag, London, UK, 2014, p.177.
[10] I. Zabaleta, E. Bizkarguenaga, D. Bilbao, N. Etxebarria, A. Prieto and O. Zuloaga, *Talanta*, 2016, 152, 353.
[11] T.H. Begley, W. Hsu, G. Noonan and G. Diachenko, *Food Additives & Contaminants: Part A*, 2008, 25, 384.
[12] C.R. Powley, M.J. Michalczyk, M.A. Kaiser and W. Buxton, *Analyst*, 2005, 130, 1299.
[13] M. Surma, W. Wiczkowski, H. Zieliński and E. Cieślik, *Packaging Technology and Science*, 2015, 28, 789.
[14] I. Zabaleta, N. Negreira, E. Bizkarguenaga, A. Prieto, A. Covaci and O. Zuloaga, *Food Chemistry*, 2017, 230, 497.
[15] M. Aznar, E. Canellas and C. Nerin, *Journal of Chromatography A*, 2009, 1216, 5176.
[16] R. Sendón, J. Bustos, J.J. Sanchez, P. Paseiro and M.E. Cirugeda, *Food Additives & Contaminants: Part A*, 2010, 27, 107.
[17] *Monograph on the Evaluation of Carcinogenic Risks to Humans*, International Agency for Research on Cancer (IARC), Lyon, France. http://monographs.iarc.fr/
[18] Y. Sanchis, C. Coscollà, M. Roca and V. Yusà, *Talanta*, 2015, 138, 290.
[19] Rapid Alert System for Food and Feed. https://webgate.ec.europa.eu/rasff-window/portal/?event = SearchForm&cleanSearch = 1
[20] EFSA Panel on Contaminants in the Food Chain (CONTAM), *Scientific Opinion on Mineral Oil Hydrocarbons in Food* in *EFSA Journal*, 2012, 10, 2704.

[21] M. Biedermann and K. Grob, *Journal of Chromatography A*, 2012, 1255, 76.

[22] European Commission Regulation (EU) No 2017/84 of 16th January 2017 on Monitoring of Mineral Oil Hydrocarbons in Food and in Materials and Articles Intended to Come into Contact with Food, *Official Journal of the European Union L*, 2017, 12, 95.

[23] J.C. Hannon, J. Kerry, M. Cruz-Romero, M. Morris and E. Cummins, *Trends in Food Science & Technology*, 2015, 43, 43.

[24] M. Jokar, G.A. Pedersen and K. Loeschner, *Food Additives & Contaminants: Part A*, 2017, 34, 434.

[25] G.O. Noonan, A.J. Whelton, D. Carlander and T.V. Duncan, *Comprehensive Reviews in Food Science and Food Safety*, 2014, 13, 679.

[26] A. Störmer, J. Bott, D. Kemmer and R. Franz, *Trends in Food Science & Technology* 2017, 63, 39.

7 Non-intentionally added substances

7.1 Chemical aspects and analysis of nonintentionally added substances

Food contact materials (FCM) can release substances into food and beverages to which the consumer will be exposed to through their diet. These substances can be intentionally added substances (IAS) or nonintentionally added substances (NIAS). The latter may be present in the final product as impurities of the monomers, additives or starting substances, as contaminants, or could be degradation or reaction products. In this context, the Commission Regulation (EU) No. 10/2011 on plastic materials and articles intended to come into contact with food establishes that "any potential health risk in the final material or article arising from reaction and degradation products should be assessed by the manufacturer in accordance with internationally recognised scientific principles on risk assessment."

Particular attention has been paid to NIAS with a molecular weight (MW) below 1,000 Da, as it is generally recognised that components with a molecular mass above 1,000 Da are not absorbed by the gastrointestinal tract and therefore should not represent a risk to consumer health.

In addition to plastic, NIAS have also been identified in other FCM such as paper and cardboard, adhesives, inks and so on.

The determination of NIAS (identification and quantification) is an extremely complex task; this is due to, among other reasons, the fact that in most cases the composition of the formulations used in the manufacture of FCM is not fully known or because of the complexity of the packaging (e.g., multilayer materials), as well as considerations relating to chemical analysis.

Advanced analytical techniques, namely, gas chromatography (GC)–mass spectrometry (MS) for volatile and semivolatile compounds, and liquid chromatography (LC) coupled to MS for nonvolatile compounds have been applied to address the challenge of identifying NIAS [1–4]. Figure 7.1 shows a scheme of an analytical procedure to evaluate NIAS.

Several research papers have been devoted to the development of analytical methodologies for the identification of NIAS in materials intended to be in contact with food. New approaches for the nontarget analysis of packaging materials have also been investigated. Here is a summary of some examples found in the scientific literature.

https://doi.org/10.1515/9783110648065-007

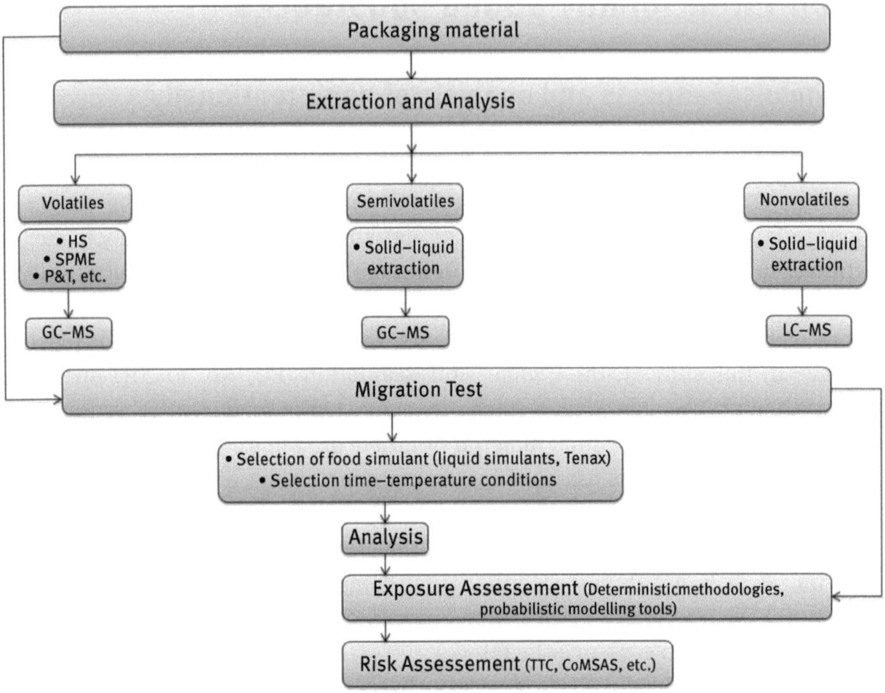

Figure 7.1: Scheme of an analytical procedure to evaluate NIAS (CoMSAS: complex mixture safety assessment strategy; HS: headspace; P&T: purge and trap; SPME: solid-phase microextraction and TTC: threshold of toxicological concern).

7.1.1 Plastic materials

7.1.1.1 Polyethylene, polypropylene, ethylene vinyl alcohol, polyethylene terephthalate and polyvinyl chloride

In a recent application, the combination of GC coupled to time-of-flight (ToF), with an electron ionisation source, and hybrid quadrupole time-of-flight (Q-ToF) mass analysers, with an atmospheric pressure chemical ionisation source, was explored for the nontarget analysis of multilayer trays made of polypropylene (PP)/ethylene vinyl alcohol (EVOH)/PP and a PP/aluminium foil/PP multilayer film. Samples were subjected to a migration test with Tenax and isooctane. The chromatographic separation was performed on a fused silica HP-5 MS capillary column (30 m × 0.25 mm ID and a film thickness of 0.25 μm). Bis(2-ethylhexyl)phthalate and dibutyl phthalate were found in both simulants, which are compounds commonly used as plasticisers that are included in the European Union (EU) positive list along with their specific migration limit. However, most of the compounds detected were nonregulated substances such as 2,4-di-tert-butyl-phenol and 2,6-di-tert-butyl-*p*-benzoquinone, which have been described as degradation products of the antioxidants Irgafos® 168 and Irganox® 1010, and have been identified as NIAS in different works. Other compounds identified were *p*-tolyl disulphide and diethyl disulphide, which are a rubber accelerator and a by-product during the production of ethanethiol, which is used as a starting and intermediate compound in plastics production, respectively [5].

Bach et al. [6] used GC coupled with ion-trap MS to determine IAS and NIAS released from polyethylene terephthalate (PET) bottles into water, produced as a result of the effect of temperature. A Rxi-5ms column (30 m × 0.25 mm ID; 0.25 µm) with a deactivated precolumn (5 m × 0.53 mm) was used as the stationary phase. Among the compounds identified, the concentration of 2,4-di-tert-butylphenol increased two-fold due to the effect of temperature, and bis(2-hydroxyethyl)terephthalate was also identified as an NIAS.

In another example reported by Panseri et al. [7], polyvinyl chloride (PVC) and polyethylene (PE) films for wrapping cheeses were analysed using HS-SPME and GC–MS to determine the presence of volatile organic compounds. A divinylbenzene–carbovax–polydimethylsiloxane fibre showed the best results and was chosen for further analysis. The chromatographic separation was performed on a Rtx-Wax column (30 m × 0.25 mm ID; 0.25 µm film thickness, Restek, Bellefonte, PA, USA). Different classes of compounds such as hydrocarbons, aldehydes, ketones, alcohols, esters and so on were identified in the material samples. For instance, 2-ethylhexanol and triacetin were found in PVC-based films but were not found in PE films. The authors noted that these compounds can probably be considered as NIAS. Cheese samples wrapped with the plastic films were also analysed and 2-ethylhexanol was detected in the cheese samples. Triacetin and ethylbenzene were also found in wrapped cheese samples, but not detected in unwrapped ones.

An in-depth study devoted to the nontarget analysis of migrants from FCM, including polyolefin (PO) drinking bottles, water boilers, polyamide (PA) cooking utensils and plastic multilayer materials, was carried out by the Norwegian Food Safety Authority for surveillance purposes [8]. PO bottles, comprising low-density PE, high-density PE and PP, were exposed to water under different time–temperature conditions to evaluate the migration. The analysis of volatile compounds was performed using P&T and GC–MS and semivolatile compounds were determined by solid-phase extraction GC–MS. Substances included in the EU positive list such as benzophenone ethyl benzoate and caprolactam were identified; however, other compounds not included were also detected, for example, 2,4-di-tert-butylphenol and 2,6-di-tert-butylbenzoquinone, which are degradation products of Irgafos® 168 and Irganox® 1076, or 7,9-di-tert-butyl-1-oxaspiro[4.5]deca-6,9-diene-2,8-dione, which could arise due to the degradation of the Irganox®-type stabiliser. With regard to plastic kettles, 2,4-di-tert-butylphenol was the prevailing migrant and its concentration increased upon successive exposure to water. Aniline and diphenylamine were detected in the migration from PA cooking utensils, and the latter has been reported as an impurity or a degradation product of the stabilisers in PA.

Derivatives of pyridine with low MW have also been identified as degradation products of PA; cyclopentanone derivatives are thermal degradation products that originate during polymerisation. Plastic laminates are complex materials that contain adhesives, printing inks and so on, among others; hence, the potential migrants include a large variety of compounds, and in the samples analysed, *N*-ethyl-toluenesulfonamide was the predominating migrant.

7.1.1.2 Polyamide

PA is a polymeric material commonly employed in articles and packaging intended to be in contact with food. PA can be synthesised from amino carboxylic acids or lactams as a starting substance, or another type of PA can be obtained by means of the condensation of diamines and dicarboxylic acids. PA6, obtained by hydrolytic or anionic polymerisation, and PA66 synthesised via the polycondensation of 1,6-diaminohexane and adipic acid are some of the PA frequently used in FCM [9]. During the synthesis of PA, cyclic oligomers are formed that can migrate into the food, and so it is necessary to evaluate them. Heimrich et al. [9] developed a method based on high-performance liquid chromatography (HPLC) coupled with a chemiluminescence nitrogen detector (CLND) to determine cyclic oligomers in PA used in FCM. Different PA-based granulates were analysed, namely, PA6, PA6/PA66, PA MXD6 and a PA6I/6T copolymer. First, the materials were submitted to dissolution/precipitation with trifluoroethanol and methanol before analysis. The chromatographic conditions were as follows: a reversed-stationary phase (RP18) and a gradient elution system consisted of 0.1% formic acid in water and methanol. Cyclic oligomers of up to $n = 9$ and $n = 14$ were detected by HPLC–CLND in PA6 and PA6/PA66 granulates, respectively, and for higher oligomers LC coupled to MS was required. In the case of PA MXD6 and PA6I/6T since they are semiaromatic PA, cyclic oligomers can also be identified by HPLC–ultraviolet (UV) (λ 260 nm and λ 250 nm).

7.1.1.3 Polycarbonate and Others

In the study reported by Simoneau et al. [10], several materials used as substitutes of polycarbonate (PC) in the manufacture of baby bottles were analysed with the aim of identifying potential migrants. The materials evaluated were PA, polyether sulfone (PES), PP, silicone, a copolyester (commercially named Tritan™) and PC baby bottles were also analysed. Migration assays were performed under hot-fill conditions (2 h, 70 °C) and using the food simulant indicated for milk, that is, 50% ethanol (v/v). Among the substances identified in PP bottles, some are not included in the European Community positive list as they are NIAS, such as the case of 2,4-di-tert-butylphenol, which could be a degradation product of Irgafos® 168; other substances not included in the positive list and that the authors also identified were 2-propenamide, 2-methyl-*N*-phenyl and *N*-butylbenzenesulfonamide. Cyclododecene was identified in PA bottles, and it has been reported that this compound could be an intermediate in the synthesis of monomers used in the production of PA, co-PA and polyesters. In some samples, Bisphenol A (BPA) was unexpectedly found, although it was branded as 'no BPA'. On the other hand, no significant migration was observed in the bottles made of Tritan™ or PES.

More recently, Onghena et al. [11] also evaluated the migration from plastic baby bottles that have replaced those of PC, specifically PP, PES, PA, Tritan™, silicone and stainless steel. GC coupled to MS triple quadrupole and a DB-5ms (30 m × 0.25 mm × 0.25 µm) column was used to perform the analyses. Ethanol (50% v/v) was also selected as the food simulant, and the migrants were subsequently extracted using ethyl acetate–*n*-hexane (1:1). Baby bottles were sterilised prior to the migration assays (2 h at 70 °C). A huge variety of compounds were identified in the materials analysed, and as per the study reported by Simoneau et al. [10], 2,4-di-tert-butylphenol and 2,6-di-tert-butylbenzoquinone were found, as well other substances not present on the positive list such as 2-butoxyethyl acetate, acetophenone and naphthalene. PES was the material with a low migration rate. In a later work [12], the authors studied the migration under 'real-life' conditions, that is, using a microwave, sterilisation and dishwasher. The authors concluded that after continued use of the baby bottles under these 'real' conditions, the migration of relevant substances would not increase.

Bignardi et al. [13] applied ultra high-performance liquid chromatography (UHPLC) and electrospray ionisation (ESI) quadrupole Orbitrap™ high-resolution mass spectrometry (HRMS) for targeted and untargeted analysis of PC-based FCM. Additives such as antioxidants and UV absorbers were selected as target compounds, a C_{18} Acclaim™ PepMap™ rapid separation liquid chromatography (150 mm × 0.3 mm, 2.0 µm particle size) column and a binary mobile phase consisting of 1 mM ammonium formate in methanol:water (10:90 v/v) and 1 mM ammonium formate in methanol were used in the analysis. Additionally, a screening was also carried out with the aim of identifying NIAS, PC degradation products and organic colourants. With respect to the latter, all the compounds found were in the list of dyes frequently used in the manufacture of FCM and are classified as "solvent colourants"; however, several oligomers of PC were also detected.

7.1.1.4 Polyester

Polyester materials are currently undergoing extensive research as the use of this polymer in FCM is increasing and it is expected that the number of applications in this field will grow.

Brenz et al. [14] proposed a method that involved alkaline hydrolysis in an organic solvent, followed by the analysis of monomers from polyester-based FCM. Polyester acids were analysed by HPLC–diode array detection (DAD) using a C_{18}-ether column and a gradient composed of 0.1% (v/v) trifluoroacetic acid in water and acetonitrile (ACN); the selected wavelengths were 210, 220, 240 and 275 nm. However, polyester polyols were determined by GC–MS and the compounds were separated on a polyethylene glycol, Optima WAXplus® (30 m × 0.25 mm ID and 0.5 μm) column. The polyester monomer composition varies depending of several factors including the type of FCM, for example, in PET samples, some of the compounds found were isophthalic acid, diethylene glycol and 1,4-cyclohexanedimethanol. In some cases, diethylene glycol can be considered as an NIAS when it is not intentionally added. Conversely, terephthalic acid and 1,4-butanediol were determined in polybutylene terephthalate samples. Regarding polyester-based resins for coatings, terephthalic acid and isophthalic acid were the main acids, with trimellitic acid as the crosslinking agent; furthermore, very low amounts of phthalic acid could be present, although the authors pointed out that, as described elsewhere, this compound could be an impurity of terephthalic acid, in which case it would be an NIAS. In another resin that was analysed, the authors also found adipic acid. With respect to polyols, there were large differences between the resins analysed; some of the substances identified include neopentyl glycol, ethylene glycol, 2-methyl-1,3-propanediol and diethylene glycol.

In a different study, several techniques, including GC–MS, HPLC–DAD/MS, HPLC–DAD/charged aerosol detector and UHPLC tandem HRMS, were applied to investigate unknown compounds from polyester can coatings. Samples were extracted with ACN and with ethyl acetate under different time–temperature conditions and subsequently analysed using the above-mentioned techniques. A total of 29 nonvolatile oligomers were tentatively identified, and the results suggest that isophthalic acid, terephthalicacid and nadic acid are the base monomers of the polyester used for the coating. The authors reported direct analysis in real-time HRMS as a powerful technique to analyse the samples directly, and applied it successfully to determine the previously identified oligomers [15]. In a later work, the migration of the monomers and oligomers, tentatively identified from the can coatings, into different food simulants [water, 3% acetic acid (w/v), 10% ethanol (v/v), 50% ethanol (v/v), and isooctane] was monitored for a long time period (515 days). The results reported an increase of the concentration of several migrants over time, particularly in ethanolic food simulants [16].

7.1.1.5 Epoxy, acrylic phenolic and polyvinyl chloride coatings

A similar approach was applied to evaluate the migration from epoxy and acrylic phenolic coatings into food simulants. BPA, Bisphenol A diglycidyl ether (BADGE), BADGE derivatives and benzoguanamine were identified as potential migrants. Additionally, two unknown compounds were also selected to be monitored during the migration assay. The authors found that in the 50% ethanol (v/v) simulant, the concentration of migrants increased after the 10th day of testing, that is, the migration of compounds continued after the conventional test period [17].

A novel method to explore potential migrants from PVC-coated cans was developed by Vaclavikova et al. [18], who used UHPLC–quadrupole Orbitrap^{TM}/MS. Several compounds such as benzoguanamine, oleamide, erucamide, acetyl tributyl citrate (ATBC), phthalates and adipates, among others, were identified and their migration into food simulants evaluated. The results showed that in the two simulants tested [water, 3% acetic acid (w/v)] the concentration of benzoguanamine increased during the migration test, which was performed for a long period (1.5 years).

7.1.2 Biodegradable packaging

Over the past few years, the use of biodegradable materials has gradually been replacing petroleum derivatives due to environmental concerns.

In an example reported in the literature, the identification and quantification of NIAS from a biodegradable adhesive has been illustrated [19], using UPLC–ESI–Q-TOF–MS^E and GC–MS. A reversed-phase C_{18} column and a gradient of methanol were the conditions selected for the UPLC analysis and a HP-5MS column (30 m × 0.25 µm × 250 µm) for the GC analysis. The migration tests were performed with simulant 'E' [poly(2,6-diphenyl-*p*-phenylene oxide) (Tenax)]. The GC–MS results revealed the presence of 1,6-dioxacyclododecane-7,12-dione, which is reported elsewhere as a degradation product in resins and could be considered as an NIAS. Regarding nonvolatile compounds, four unknown compounds were detected, 1, 6,13,18-tetraoxacyclotetracosane-7,12,19,24-tetraone, 1,6,13,18,25,30-hexaoxacyclohexatriacontane-7,12,19,24,31,36-hexone, 1,6,13,18,25,30,37,42-octaoxacyclooctatetracontane-7,12,19,24,31,36,43,48-octaone and 1,6,13,18,25,30,37,42,49, 54-decaoxacy clohexacontane-7,12,19,24,31,36,43,48,55,60-decaone. The authorspointed out that these compounds and 6-dioxacyclododecane-7,12-dione could be neo-formed compounds as a result of the interaction of some volatile and nonvolatile compounds identified in the material. Two of these neo-formed compounds, 1,6-dioxacyclododecane-7,12-dione and 1,6,13,18-tetraoxacyclotetracosane-7,12,19,24-tetraone, and an IAS, 2,4,7,9-tetramethyl-5-decyne-4,7-diol, migrated into Tenax.

7.1.3 Printing inks

Printing inks are commonly employed to print the external face of food packaging. Several studies have reported the migration of ink components, such as photoinitiators, by direct contact, through the vapour phase or by the set-off phenomenon [20–22]. Besides the known substances, formulations may also contain NIAS, such as impurities, reaction products and so on, which should be evaluated.

In a recent study, Aznar et al. [23] investigated the migration of components from multilayer materials containing inks into two food simulants, Tenax and 95%

ethanol (v/v). UPLC–Q-TOF–MS was used to perform the analysis and the compounds were separated on a C18 column (2.1 × 100 mm, 1.7 μm particle size) using a binary mobile phase consisting of water with 0.1% formic acid (A) and methanol with 0.1% formic acid (B). MS data were acquired in the positive ion ESI mode. The composition of multilayer materials evaluated in the work comprised PET, ink, adhesive, aluminium, gloss lacquer, matt lacquer and PE, and the other material comprised oriented PP, aluminium, adhesive, ink, paper, matt lacquer and PE. Several migrants from the ink were identified, for instance, plasticisers such as sebacates, ATBC and so on, glycol ethers, cyclic polyesters such as 1,4,7-trioxacyclotridecane-8,13-dione, which has been reported as an NIAS in materials produced with polyurethane (PU) adhesives. The authors also evaluated the effect of lacquer on the migration of ink components by the set-off phenomenon and, in general, they found that the lacquer reduced the transference of substances. Regarding the paper-based multilayer materials, some of the migrants detected were the same as those identified in the other materials, such as ATBC, cyclic polyester and so on; however, three new compounds were also identified, that is, oleamide, stearamide and behenamide. Furthermore, the effect of adding lacquer resulted in a reduction of the migration of some compounds but an increase of the migration of other compounds.

7.1.4 Adhesives

The components and potential migrants, including NIAS, from adhesives have been extensively investigated. Several scientific papers have been devoted to identifying the presence of NIAS in adhesives.

In a study reported by Félix et al. [24], HS-SPME–GC–MS together with ChemSpider and SciFinder databases were used for the identification of NIAS from different samples of PU adhesives in multilayer materials; plastic films including PET, PA, PP, PE and PE/ethyl vinyl alcohol were also analysed. In addition, migration tests using Tenax and isooctane as food simulants were also performed. The authors reported the identification of more than 63 volatile and semivolatile compounds. Among them, two NIAS were identified, 1,6-dioxacyclododecane-7,12-dione and 1,4,7-trioxacyclotridecane-8,13-dione. The former was found in different samples of PU adhesives and, as described in the paper, it was formed by the cyclisation of the degradation products of the alkyl part; the latter was detected in PU adhesives as well as in the plastic films of PET and oriented PA. Both compounds migrated into the two food simulants.

Pezo et al. [25] identified up to 40 NIAS using Q-TOF/MS^E in different multilayer packaging containing several PU adhesive formulations. Some examples of the NIAS identified were 1,8,15-triazacycloheneicosane-2,9,16-trione, 1,3-bis(isocyanatomethyl)-cyclohexane, 1,4,7,18,21-pentaoxa-11,14,25,28-tetraazacyclohentriacontane,

1,8,15,22-tetraazacyclooctacosane-2,9,16,23-tetrone, 1-cyanodecane and caprolactam cyclic hexamer, among others. It is interesting to point out that the NIAS identified in the materials were also found in the 3% (w/v) acetic acid food simulant after the migration assay.

In a work conducted by Canellas et al. [26], the atmospheric pressure gas chromatography (APGC) coupled to Q-TOF/MS technique was investigated as an alternative and powerful tool to identify NIAS. The authors applied this technology to successfully analyse acrylic adhesives used in food packaging materials.

Later, the same authors, Canellas et al. [27], published a study in which they used the APGC–Q-TOF/MS technique to identify volatile migrants in pressure-sensitive adhesives, which is applied as autoadhesive labels and used for direct contact with food. Among the migrants detected in the samples, an NIAS was identified, methyl-3-(3,5-di-tert-butyl-4-hydroxyphenyl)propanoate, which is a degradation product of the antioxidant Irganox® 1076 and is commonly used in adhesive formulations. Migration studies were performed to test the potential migration of the chemicals into food. For that purpose, two food simulants were selected, Tenax and isooctane. Since these types of autoadhesive labels are, in general, employed for sausages or cured meat, natural pork intestine filled with isooctane was used to carry out the migration test in order to mimic real conditions. The results obtained with Tenax led to underestimation of the migration, while those obtained with isooctane as the simulant led to an overestimation, leading the authors to conclude that it would be interesting to conduct the migration assays in real food samples.

In another study carried out by Isella et al. [28], the nonvolatile compounds present in PU adhesives and potential migrants were studied. UPLC–Q-TOF/MS and the ChemSpider database were used for identification purposes. Methanol extracts of the adhesive and the antistatic additive were analysed, and the compounds were separated on an Acquity UPLC ethylene-bridged hybrid C_{18} column of (100 mm × 2.1 mm, 1.7 μm particle size). Several NIAS including 1,6-dioxacyclododecane-7,12-dione dimer, 1,4-dioxacyclotridecane-5,13-dione, 1,4,14,19-tetraoxacyclopentacosene-5,13,20,25-tetraone, 1,4-dioxacyclotridecane-5,13-dione, bis[2-(diethylamino)ethyl] 4,4'-[(2-methyl-1,3-propanediyl)bis(oxycarbonylimino)] dibenzoate and bis[2-(diethylamino)ethyl] 4,4'-[1,5-pentanediylbis(oxycarbonylimino)] dibenzoate were identified, among others. The latter compounds could be degradation products or by-products formed during the curing process. The migration assays were carried out in Tenax and the results revealed that some of the compounds previously identified had migrated into the food stimulant.

UPLC–ESI–TOF–MS and UPLC–ESI–high-definition mass spectrometry (HDMS) were the techniques selected by Canellas et al. [29] to investigate potential nonvolatile migrants from acrylic adhesives. The adhesives were extracted with ACN and subsequently, the resulting extracts were analysed by UPLC using a C_{18} column (2.1 mm × 100 mm, 1.7 μm particle size) as a stationary phase. During the first step, a TOF mass spectrometer with ESI in a positive and negative mode were applied;

in the second step, to identify single compounds and to obtain the fragmentation of each single mass for structural elucidation, HDMS and the MassFragment™ tool were used. Among the compounds identified were nonionic surfactants such as octyl phenol ethoxylate and nonylphenol ethoxylate (NPEO). However, in the case of NPEO, only the small polymers with low masses migrated into Tenax.

In a different work, Aznar et al. [30] studied the composition of different types of adhesives employed in food packaging using HS-SPME–GC–MS as the analytical technique and Tenax as the food simulant for migration tests. The samples used in the study included PU, acrylics and vinyl adhesives, among others. Fifty-five different compounds were detected and 33 identified, such as isocyanates, like toluene-2,6-diisocyanate, in PU adhesives, phthalates, like dibutyl phthalate and 2-ethylhexyl phthalate, in vinyl adhesives or abietic acid derivatives in hot-melt adhesives. Other compounds like butylated hydroxytoluene (BHT) were found in PU, vinyl and hot-melt adhesives.

7.1.5 Paper and board

Paper and cardboard are usually employed in food-packaging materials either as primary, secondary or tertiary packaging. In a recent work, Guazzotti et al. [31] performed a screening analysis using GC–MS in order to identify potential migrants in several samples of paper and board including packaging for takeaway foods and also primary and secondary packaging. Prior to GC–MS analysis, samples were extracted with a mixture of ethanol–hexane (1:1) for 2 h at room temperature. The chromatographic separation was performed on a DB-5ms (30 m × 0.25 mm ID, film thickness 0.25 μm) column. The compounds identified were classified into six groups: aliphatic and cyclic hydrocarbons, aromatics, aldehydes, ketones and ethers, alcohols and others, with aromatics being the predominant compounds in the samples analysed. Residues of solvents such as toluene and plasticisers such as diisobutyl phthalate and diisopropylnaphthalenes were found in corrugated and printed packaging, whereas triacetin was the main compound in plastic-laminated packaging materials.

A novel approach for the safety assessment of unknown NIAS from carton samples, some of which were intended to come into contact with food, has been reported by Koster et al. [32]. The valuable tool (CoMSAS) allows discrimination between toxicologically relevant and less relevant substances, at low exposure levels. This approach uses an exposure threshold of 90 μg/day, instead of the conventional limit of detection of 10 μg/kg for migrants from FCM, and substances above this value must be identified. Several techniques were used for the screening analysis, namely, GC–MS for volatile and semivolatile compounds, and to analyse nonvolatiles by GC–MS a derivatisation step was carried out with *N*-methyl-*N*-(trimethylsily)trifluoroacetamide, and an LC–UV–evaporative light-scattering detector (ELSD) was used for nonvolatile substances. Migration tests were conducted with Tenax as the food stimulant at 60 °C for 10 days. After the migration assay, Tenax was extracted with ether. An AT-5MS column (30 m × 0.25 mm, 1 µm) was used for volatile compounds, whereas for semivolatile compounds the column of choice was an AT-5MS column (30 m × 0.25 mm, 0.25 µm). For nonvolatile and semivolatile substances, a HP-5 MS VI (30 m × 0.25 mm × 0.25 µm) column was selected. The LC–UV–ELSD method involved the use of Atlantis T3 (150 × 3.0, 5 µm) and a gradient consisting of A) Milli-Q® water with 0.1% formic acid and B) ACN with 0.1% formic acid as the stationary phase and mobile phase, respectively. All volatile compounds detected in the carton were below 90 µg/day. Six semivolatile substances and about 40 nonvolatile (derivatised) substances above 90 µg/day were found, and of the nonvolatile substances detected by LC–ELSD, none were above 90 µg/day; however, mineral oils were detected in the underivatised extract.

7.1.6 Silicone materials

Nowadays, silicone articles are being used in different food contact applications. In a work reported by Helling et al. [33], the migration of different compounds from silicone materials was illustrated through two example moulds for pizza and baby teats after long-term and repeated use. The silicone moulds were subjected to repeated use under regular conditions of use in a pizza bakery and baby teats were sterilised repeatedly via microwave radiation. The main migrants from silicone materials were siloxane oligomers; with respect to volatile organic compounds (VOC), the results revealed that successive use did not produce new low and midsize volatile cyclic oligomers and it was expected that with use, the migration would decrease. In the case of baby teats, a slight reduction of the VOC released was observed after sterilisation. In another assay, the authors studied the extractable substances, which were determined and characterised by ^{1}H-nuclear magnetic resonance and chromatographic techniques (GC–MS, HPLC–ELSD, HPLC–MS). After continued use, a reduction of extractable siloxanes was observed along with an increase of fat incorporated into the mould as a result of contact with the food. However, no important changes of the amount of extractable siloxanes were observed throughout the assay with the baby teats, but augmented fat incorporation was observed. The authors note that with respect to migration, silicone moulds for long-term use are not a cause for concern.

7.1.7 Irradiated food packaging

Food irradiation has recently been applied as a preservation technique to extend the shelf life of products. As it is sometimes applied to packed food, the process may cause changes in packaging resulting in the migration of radiolytic products (NIAS) into food; hence, it is necessary to check the safety of irradiated packaging. In a study reported by Driffield et al. [34], analytical screenings were carried out by HS–GC–MS and LC–TOF–MS in different samples of packaging materials before and after irradiation. Irradiation was performed using a gamma and electron beam. An enormous number of compounds were detected by GC–MS; some examples of the substances formed after irradiation include 1,3-di-tert-butylbenzene, 2,4-di-tert-butylphenol, acetaldehyde and so on. Concerning LC–TOF–MS analysis, three of the detected compounds were reported as oxidised oligomers of polyadipate plasticisers. The authors concluded that at the same dose there were no important differences between the two types of irradiation with regards to radiolytic product formed, although, in general, the concentration of products was higher with gamma irradiation.

In a more recent report, Salafranca et al. [35] evaluated the effect of oxygen and long-term storage on the volatile compounds liberated from polymeric multilayer FCM sterilised by gamma irradiation. SPME–GC–MS was the technique of choice to determine the compounds. In nonirradiated samples, the compounds commonly identified were alkanes, alkenes and some alcohols, and no important differences were observed depending on the filling gas used. On the contrary, in irradiated samples an intense peak corresponding to 1,3-di-tert-butylbenzene was detected. Another remarkable fact is that the antioxidant BHT, present in nonirradiated samples, disappeared with its oxidation products [2,6-di-tert-butyl-1,4-benzoquinone, 2,4-di-tert-butylphenol or 7,9-di-tert-butyl-1-oxaspiro(4,5) deca-6,9-dien-2,8-dione] found in the samples after treatment. The main oxidation products in treated materials were short chain acids. With regard to the period of storage, no significant qualitative differences were observed when the samples were stored.

Tyapkova et al. [36] proposed high-resolution GC with olfactometry to determine the off-odour originating from gamma irradiation of PP. A great number of compounds were identified in both nonirradiated and irradiated samples. Some examples of compounds formed during irradiation and detected in the samples were acetic acid, 2,3-pentanedione, pentanoic acid, 3-methylbutanoic acid and so on, and these compounds have been reported in previous studies. Additionally, new compounds reported for the first time, in both types of samples, were also identified such as (*Z*)-2-nonenal, (*E*)-2-nonenal, (*E,Z*)-2,6-nonadienal, 2-methylpentanoic acid, 2-methylhexanoic acid and so on.

7.1.8 Active food packaging

Active packaging is being used to improve the quality and increase the shelf life of packaged products by releasing active compounds such as antioxidants or antimicrobials, or by removing undesired compounds (i.e., ethylene, oxygen and so on) from the environment [37, 38]. Like other packaging materials, their safety must be evaluated so the identification of potential unknown migrants is crucial.

In a study reported by Aznar et al. [39], a novel technology, UPLC–Q-TOF/MS, was applied to investigate nonvolatile compounds in new antimicrobial active packaging. The films studied were based on PP, an EVOH copolymer and PET. Migration tests were performed using 10% ethanol (v/v) and 95% ethanol (v/v) as food simulants. With respect to the chromatographic conditions, a C_{18} (2.1 × 100 mm, 1.7 µm particle size) column and a mobile phase consisting of water with 0.1% formic acid and methanol with 0.1% formic acid was employed. Besides the active compounds, NIAS were also examined. The study illustrated that impurities or degradation products of the active agents can also migrate, and so it is necessary to identify these unknown compounds. For example, in PP/EVOH films containing citral, a possible thermal reaction product of this compound was detected in the food simulant and it was identified as 3,7-dimethyl-2,3-hydroxy-6-octenal.

In another work, the authors also used MS detection with a Q-TOF analyser to study the unknown compounds in an active packaging based on PET and ethyl-*N*-dodecanoyl-L-arginate hydrochloride as the active agent. The food simulants selected for the migration test were the same used in the previous work, and in addition, the assay was performed in chicken. Some of the compounds identified were, for instance, *N*-2-dodecanoyl-L-arginine, which has been reported to be a metabolism degradation compound of ethyl lauroyl arginate, or dipropylene glycol methyl ether, which came from the coating. Both compounds were detected in chicken samples wrapped with the active films, in addition to ethyl lauroyl arginate [40].

A novel approach for the identification of NIAS in antimicrobial nanofilms for food packaging applications has been reported by Martínez-Bueno et al. [41]. Active films were prepared with polylactic acid, polylimonene and zinc oxide (ZnO) nanoparticles (NP) as the antimicrobial agents. The migration assays were carried out in food simulants A [10% ethanol (v/v)] and B [3% acetic acid (w/v)]. To conduct the study, GC and LC coupled to quadrupole Orbitrap™ mass spectrometers was used; in addition, inductively coupled plasma–MS was employed to determine ZnO NP. The liquid chromatographic conditions were as follows: a C_8 column and a binary gradient system based on water, methanol and 5 mM of ammonium formate and 0.1% formic acid. For GC analysis a TG–OCP I (trace gold organochlorine pesticides) column (30 m × 0.25 mm ID, 0.25 µm) was used. Three volatile (tripropylene glycol diacrylate, 10-heneicosene and α-tocopherol acetate) and four nonvolatile {*N,N*-diethyldodecanamide, *N*-[(9Z)-9-octadecen-1-yl]acetamide, 1-palmitoylglycerol and glycerol stearate} NIAS were identified. Higher concentrations of ZnO NP were found in the 3% acetic acid (w/v) food simulant.

7.2 Risk assessment of nonintentionally added substances

To evaluate the safety of FCM, it is necessary to know the migration of potential compounds from the packaging, which includes substances added intentionally during the manufacture of the materials and NIAS. How to approach the risk assessment of NIAS present in FCM is still under discussion. Different methodologies have been used for the risk assessment of FCM and the TTC concept has been applied for this purpose. This approach is a screening tool that is used when the toxicity of a substance is unknown or the information is very limited. To apply this tool, it is necessary to know the chemical structure of the molecule and data regarding human exposure. It employs the Cramer decision tree that classifies substances according to their molecular structure into three classes, namely, Class I, II and III for substances with low, intermediate and high toxicity, respectively. The threshold for Cramer Class I, II and III are 1,800, 540 and 90 μg/person/day, respectively [1, 42–46].

The CoMSAS is another interesting methodology developed by Koster et al. [32]. In this approach, an exposure threshold of 90 μg/day is used, which corresponds to the threshold defined for Cramer Class III, whereas the conventional limit of detection of 10 μg/kg is used for migrants from FCM and substances over this value have to be identified. One benefit of this approach, from an analytical point of view, is that the identification can be simplified. Nevertheless, this method supposes that all the substances present a similar response with different detectors.

In the near future, new tools will be explored for the risk assessment of unknown substances.

References

[1] C. Nerin, P. Alfaro, M. Aznar and C. Domeño, *Analytica Chimica Acta*, 2013, 775, 14.

[2] *Note for Guidance for Petitioners Presenting an Application for the Safety Assessment of a Substance to be used in Food Contact Materials Prior to its Authorisation*, European Food Safety Agency, 31st July 2008. http://www.efsa.europa.eu/en/efsajournal/doc/21r.pdf

[3] European Commission Regulation (EU) No 10/2011 of 14th January 2011 on Plastic Materials and Articles Intended to Come into Contact with Food, *Official Journal of the European Union L*, 2011, **12**, 1.

[4] S. Koster, M.H. Bani-Estivals, M. Bonuomo, E. Bradley, M.C. Chagnon, M.L. Garcia, F. Godts, T. Gude, R. Helling, P. Paseiro-Losada, G. Pieper, M. Rennen, T. Simat and L. Spack in *Guidance on Best Practices on the Risk Assessment of Non Intentionally Added Substances (NIAS) in Food Contact Materials and Articles*, International Life Sciences Institute Europe, 16th July 2015. http://ilsi.org/publication/guidance-on-best-practices-on-the-risk-assessment-of-non-intentionally-added-substances-nias-in-food-contact-materials-and-articles/

[5] L. Cherta, T. Portolés, E. Pitarch, J. Beltran, F.J. López, C. Calatayud, B. Company and F. Hernández, *Food Chemistry*, 2015, 188, 301.

[6] C. Bach, X. Dauchy, I. Severin, J.F. Muñoz, S. Etienne and M.C. Chagnon, *Food Chemistry*, 2013, 139, 672.

[7] S. Panseri, L.M. Chiesa, A. Zecconi, G. Soncini and I. De Noni, *Molecules*, 2014, 19, 8707.
[8] I. Skjevrak, C. Brede, I.L. Steffensen, A. Mikalsen, J. Alexander, P. Fjeldal and H. Herikstad, *Food Additives & Contaminants*, 2005, 22, 10, 1012.
[9] M. Heimrich, M. Bönsch, H. Nickl and T.J. Simat, *Food Additives & Contaminants: Part A*, 2012, 29, 5, 846.
[10] C. Simoneau, L. van den Eden and S. Valzacchi, *Food Additives & Contaminants: Part A*, 2012, 29, 3, 469.
[11] M. Onghena, E. van Hoeck, P. Vervliet, M.L. Scippo, C. Simon, J. van Loco and A. Covaci, *Food Additives & Contaminants: Part A*, 2014, 31, 12, 2090.
[12] M. Onghena, E. van Hoeck, N. Negreira, L. Quirynen, J. van Loco and A. Covaci, *Food Additives & Contaminants: Part A*, 2016, 33, 5, 893.
[13] C. Bignardi, A. Cavazza, C. Corradini and P. Salvadeo, *Journal of Chromatography A*, 2014, 1372, 133.
[14] F. Brenz, S. Linke and T. Simat, *Food Additives & Contaminants: Part A*, 2017, 34, 2, 307.
[15] R. Paseiro-Cerrato, S. MacMahon, C.D. Ridge, G.O. Noonan and T H. Begley, *Journal of Chromatography A*, 2016, 1444, 106.
[16] R. Paseiro-Cerrato, G.O. Noonan and T.H. Begley, *Journal of Agricultural and Food Chemistry*, 2016, 64, 2377.
[17] R. Paseiro-Cerrato, J. DeVries and T.H. Begley, *Journal of Agricultural and Food Chemistry*, 2017, 65, 2594.
[18] M. Vaclavikova, R. Paseiro-Cerrato, L. Vaclavika, G.O. Noonan, J. DeVries and T.H. Begley, *Food Additives & Contaminants: Part A*, 2016, 33, 2, 352.
[19] E. Canellas, P. Vera and C. Nerín, *Analytical and Bioanalytical Chemistry*, 2015, 407, 6781.
[20] M.A. Lago, A. Rodríguez-Bernaldo de Quirós, R. Sendón, J. Bustos, M.T. Nieto and P. Paseiro, *Food Additives & Contaminants: Part A*, 2015, 32, 5, 779.
[21] A. Rodríguez-Bernaldo de Quirós, R. Paseiro-Cerrato, S. Pastorelli, R. Koivikko, C. Simoneau and P. Paseiro-Losada, *Journal of Agricultural and Food Chemistry*, 2009, 57, 10211.
[22] R. Koivikko, S. Pastorelli, A. Rodríguez-Bernaldo de Quirós, R. Paseiro-Cerrato, P. Paseiro-Losada and C. Simoneau, *Food Additives & Contaminants: Part A*, 2010, 27, 10, 1478.
[23] M. Aznar, C. Domeño, C. Nerín and O. Bosetti, *Dyes and Pigments*, 2015, 114, 85.
[24] J.S. Félix, F. Isella, O. Bosetti and C. Nerín, *Analytical and Bioanalytical Chemistry*, 2012, 403, 2869.
[25] D. Pezo, M. Fedeli, O. Bosetti and C. Nerín, *Analytica Chimica Acta*, 2012, 756, 49.
[26] E. Canellas, P. Vera, C. Domeño, P. Alfaro and C. Nerín, *Journal of Chromatography A*, 2012, 1235, 141.
[27] E. Canellas, P. Vera and C. Nerín, *Journal of Mass Spectrometry*, 2014, 49, 1181.
[28] F. Isella, E. Canellas, O. Bosetti and C. Nerin, *Journal of Mass Spectrometry*, 2013, 48, 430.
[29] E. Canellas, C. Nerín, R. Moore and P. Silcock, *Analytica Chimica Acta*, 2010, 666, 62.
[30] M. Aznar, P. Vera, E. Canellas, C. Nerín, P. Mercea and A. Störmer, *Journal of Materials Chemistry*, 2011, 21, 4358.
[31] V. Guazzotti, B. Giussani, L. Piergiovanni and S. Limbo, *Packaging Technology and Science*, 2015, 28, 5, 385.
[32] S. Koster, M. Rennen, W. Leeman, G. Houben, B. Muilwijk, F. van Ackerc and L. Krula, *Food Additives & Contaminants: Part A*, 2014, 31, 3, 422.
[33] R. Helling, P. Seifried, D. Fritzsche and T.J. Simat, *Food Additives & Contaminants: Part A*, 2012, 29, 9, 1489.
[34] M. Driffield, E.L. Bradley, I. Leon, L. Lister, D.R. Speck, L. Castle and E.L.J. Potter, *Food Additives & Contaminants: Part A*, 2014, 31, 3, 556.

[35] J. Salafranca, I. Clemente, F. Isella, C. Nerín and O. Bosetti, *Analytica Chimica Acta*, 2015, 878, 118.
[36] O. Tyapkova, M. Czerny and A. Buettner, *Polymer Degradation and Stability*, 2009, 94, 757.
[37] N. de Kruijf, M. van Beest, R. Rijk, T. Sipiläinen-Malm, P. Paseiro Losada and B. De Meulenaer, *Food Additives & Contaminants*, 2002, 19, Supplement, 144.
[38] M. Imran, A.M. Revol-Junelles, A. Martyn, E.A Tehrany, M. Jacquot, M. Linder and S. Desobry, *Critical Reviews in Food Science and Nutrition*, 2010, 50, 799.
[39] M. Aznar, A. Rodriguez-Lafuente, P. Alfaro and C. Nerin, *Analytical and Bioanalytical Chemistry*, 2012, 404, 1945.
[40] M. Aznar, J. Gómez-Estaca, D. Vélez, V. Devesa and C. Nerín, *Food and Chemical Toxicology*, 2013, 56, 363.
[41] M.J. Martínez-Bueno, M.D. Hernando, S. Uclés, L. Rajski, S. Cimmino and A.R. Fernández-Alba, *Talanta*, 2017, 172, 68.
[42] W. Leeman and L. Krul, *Current Opinion in Food Science*, 2015, 6, 33.
[43] *Exploring Options for Providing Advice about Possible Human Health Risks based on the Concept of Threshold of Toxicological Concern* in *EFSA Journal*, 2012, 10, 7, 2750.
[44] *Review of the Threshold of Toxicological Concern (TTC) Approach and Development of New TTC Decision Tree*, European Food Safety Authority and the World Health Organization, EFSA Supporting Publication 2016:EN-1006. https://www.efsa.europa.eu/en/supporting/pub/1006e
[45] M.A.J. Rennen, S. Koster, C.A.M. Krul and G.F. Houben, *Food and Chemical Toxicology*, 2011, 49, 933.
[46] S. Koster, A.R. Boobis, R. Cubberley, H.M. Hollnagel, E. Richling, T. Wildemann, G. Würtzen and C.L. Galli, *Food and Chemical Toxicology*, 2011, 49, 1643.

8 Estimation of exposure to migrants from food contact materials

The estimation of exposure to chemicals through the diet is of crucial importance to the food safety field. Nowadays, the majority of foods are commercially packed, and so the food-packaging materials represent a potential source of contamination as some chemicals can migrate from the food contact material (FCM) into the food. Therefore, food-packaging materials should be under risk evaluation via dietary exposure assessments [1].

Different approaches or methodologies have been developed and applied for the exposure assessment of migrants from FCM. In the European Union (EU), a conservative and simple tool is used and is based on the following assumption: '1 kg of food is consumed daily by a person of 60 kg bodyweight (bw) and that the food is packaged in a cubic container of 6 dm^2 surface area releasing the substance' and it is also assumed that migration takes place at a maximum level [2–4]. It is interesting to point out that although, in general, this supposition represents the worst-case scenario, in some cases this is not fulfilled, for example, for children, as has been demonstrated in several studies [3, 5, 6].

Hammarling et al. [5] studied the migration of epoxidised soybean oil (ESBO) from plasticised polyvinyl chloride (PVC) gaskets into baby food and found that 5% of the samples analysed exceeded the acceptable level, taking into account the tolerable daily intake (established for adults) of 1 mg/kg bodyweight proposed by the Scientific Committee for Food of the European Commission and a bodyweight of 8.5 kg for an infant of 8 months. Later, Fantoni and Simoneau [6] conducted a European survey to determine the chemicals in baby food and they observed that in 14.9% of the samples, the level was exceeded.

More refined methods are required to achieve a more realistic estimation of consumer exposure. The approach employed by the US Food and Drug Administration (FDA) involves the use of factors. The estimation was performed by combining migration data with data regarding the use of the food contact articles in which a food contact substance may be present. As is stated in the US FDA guidance document, 'the term "consumption factor" describes the fraction of the daily diet expected to contact specific packaging materials' and the '"food-type distribution" factors are calculated for each packaging material to reflect the fraction of all food contacting each material that is aqueous, acidic, alcoholic and fatty' [7]. Detailed information about how to calculate the exposure estimation is reported in the above document. This method seems to be more realistic than that used in the EU, but it is also deterministic [8]. Hence, it is desirable to have more refined methods and probabilistic methodologies for exposure assessment have been developed. Within the Flavours, Additives and Food Contact Materials Exposure Task project (FACET), part-funded by the European Commission under the Framework FP7 Programme, a probabilistic

https://doi.org/10.1515/9783110648065-008

modelling tool was developed to determine the exposure to migrants from food-packaging materials. For a specific migrant and a defined population, the FACET tool uses data regarding the packaging that contains the migrant, the foods that can be contained in this package, determines the concentration of the migrant in the food by migration modelling and with food consumption data estimates consumer exposure. A detailed explanation about how the FACET tool was constructed employing different databases is described elsewhere [9].

The exposure assessment is an important part of risk assessment; one of the methods used to investigate human exposure to chemicals through the diet is the so-called total diet study (TDS), which is an economical and convenient approach. The main characteristics in this type of study are that it should be representative of the whole diet, and the food is analysed as consumed and grouped into pools resulting in a more realistic exposure estimation. Furthermore, this approach can be used as a screening tool or for a more refined exposure assessment [10].

TDS have been applied to evaluate the exposure to chemicals and nutrients. In a recent work reported by Vin et al. [11], an overview of the suitability of the TDS approach to different chemical substances was presented. Some of the examples within the group of chemicals that migrated from food packaging are melamine, mineral oil hydrocarbons, Bisphenol A and phthalates. The main sources of exposure to Bisphenol A (BPA) through the diet are polycarbonate (PC) materials and epoxy resins, while in the case of melamine it is via melamine-based resins and melamine formaldehyde (MF) plastics [11–14].

Phthalates are commonly used as plasticisers in the manufacture of FCM, although their use has been limited in recent years; however, there are still many materials that contain them with diet being the most important route of exposure [15–17]. Some representative examples of studies on the exposure assessment of chemicals transferred from FCM are summarised in Table 8.1.

Table. 8.1: Overview of studies about the assessment of the exposure to migrants from food contact materials.

Migrant/chemical	Source	Exposure estimation (Tool)	Exposure	Reference
Di(2-2-ethylhexyl) phthalate (DEHP)	Food contact materials	Combining the concentration of DEHP in foods with consumption data	Country: China Population mean 2–6 years – 4.51 μg/kg bw/day 7–12 years – 3.41 μg/kg bw/day 13–17 years – 2.46 μg/kg bw/day ≥ 18 years – 2.03 μg/kg bw/day general 2.34 μg/kg bw/day	[18]
Di-butyl-phthalate (DBP)	Paperboard packaging	Probabilistic analysis with a Monte Carlo sampling method	Portuguese population 0–8.95 μg/day/kg bw (mean 0.44 μg/day/kg bw)	[19]
DEHP, DBP, benzyl butyl phthalate (BBP), and diethyl phthalate (DEP)	Food contact materials	The exposure was calculated based on phthalate concentrations in foods, the food consumption frequency per day, gastrointestinal uptake efficiency and body weight	Adult Swiss-German population DEHP 1.90 μg/kgbw/day DBP 0.39 μg/kgbw/day BBP 0.14 μg/kgbw/day DEP 0.02 μg/kg bw/day	[20]

(continued)

Table. 8.1 (continued)

Migrant/chemical	Source	Exposure estimation (Tool)	Exposure	Reference
Dimethyl phthalate (DMP), DEP, butyl benzyl phthalate (BBzP), di-*n*-butyl phthalate (DBP), diisobutyl phthalate (DiBP), DEHP, di-*n*-hexyl phthalate (DnHP), dicyclohexyl phthalate (DCHP), and di-*n*-octyl phthalate (DnOP)	Food containers	EPA (US Environmental Protection Agency). Exposure Factors Handbook: 2011	Country: USA, Albany, New York Population: adult and child Values ranged from 0.004 µg/kg/day for DMP to 0.673 µg/kg/day for DEH	[21]
DEP, di-*n*-butyl phthalate (DnBP), BBP and DEHP	Environment Migration from food contact materials	Semi-probabilistic modelling approach (extended version of ENvironmental Food transfer model for ORganic Contaminants the EN-forc model)	Country: Belgium Population: adult (15 – 98 years) DEP 0.039 µg/kg bw/day (average) DnBP 0.160 µg/kg bw/day BBP 0.088 µg/kg bw/day DEHP 1.45 µg/kg bw/day	[22]
Bisphenol A (BPA)	Food packaging (cans, fruit juice bottles/packs) and microwave containers	The exposure was calculated combining the concentration of BPA in foods migrated from food packaging with consumption data	Country: Spain Population: 6- to 8-year old school children 0.025 µg/kg bw/day (mean)	[23]
BPA	Food cans and plastic microwave containers	The exposure was calculated combining the concentration of BPA in foods with consumption data	Country: Spain Population: cohort of mother–son pairs Total BPA (ng/day): 1,078.91 (mean)	[24]

BPA	Food-contact products or packages of polycarbonate resins	The exposure was estimated using the consumption factor (CF) and Food-Type Distribution Factors (Ft) (US FDA 'Recommendations for Chemistry Data for Indirect Food Additive Petitions')	< 0.0000125 mg/kg bw/day	[25]
BPA	Food and beverage cans coated with bisphenol A-based epoxies	The exposure was estimated in accordance to US FDA guidance ('Recommendations for Chemistry Data for Indirect Food Additive Petitions'), using the CF and food-type distribution factors	2.2 ppb	[26]
BPA	Polycarbonate bottles of drinking water	The exposure was estimated taking into account the highest detected level in water samples (worst-case scenario) and considering a daily consumption of 2 L of water by an adult of 60 kg of body weight and 0.75 L for an infant of 5 kg of body weight	Country: Lebanon Population: adult and infant 0.046 µg/kg/day for adults 0.20 µg/kg/day for infants	[27]

(continued)

Table. 8.1 (continued)

Migrant/chemical	Source	Exposure estimation (Tool)	Exposure	Reference
BPA	Paper and paper products including food contact papers and others (not intended to be in contact with foods)	Estimation of median daily intake of BPA through dermal absorption from based on handling or touching of paper product	Total exposure (geometric mean): 12.5 ng/day Food contact papers: 0.0006 ng/day Food cartons: 0.0019 ng/day Estimated median daily intake value: 0.219 ng/kg bw/day for the general population	[28]
BPA	Cans and all other food packaging	The exposure assessment was performed by combining concentrations of BPA in foods with data of consumption ('fraction canned')	Overall intake: 12.6 ng/kg-day Canned foods: 12.4 ng/kg-day Canned vegetables: 11.9 ng/kg-day Canned meat: 0.4 ng/kg-day	[29]
BPA	Plastic containers	The exposure was calculated taking into account the migration levels detected from the water bottles and considering an adult of 60 kg of body weight and that consumes 1.5 L of water per day and for the upper range of exposure that consumes 2 L of water per day	4.00×10^{-5} mg/kg bw/day 1.48×10^{-4} mg/kg bw/day (upper range of exposure)	[30]

BPA	Food packaging (light metal packaging)	Refined deterministic approach Fully probabilistic approach (FACET)	Country: UK, Population: 19–64 year olds 0.0098 mg/person/day (mean)	[31]
DMP, DEP, DiBP, DiDP, DiNP, DnBP, BBzP, DEHP, DCHP, DnOP and BPA	Food packaging (plastic, paper, cardboard box, metal foil, canned, metal tube, aluminium foil, glass with metal or plastic screw cap)	Diet calculation system Kost Beregnings System (KBS) version 7.0	Country: Norway Population: adult (18 – 70 years) DEHP 400–500 ng/kg bw/day DiNP 400–500 ng/kg bw/day DiBP 30–40 ng/kg bw/day DnBP 30–40 ng/kg bw/day DnOP 30–40 ng/kg bw/day DiDP 30–40 ng/kg bw/day DMP 10–20 ng/kg bw/day DEP10–20 ng/kg bw/day DCHP 10–20 ng/kg bw/day BPA 5 ng/kg bw/day	[32]
BPA and DEHP	Food packaging	Determining urinary BPA and phthalate metabolites before, during and after dietary intervention (limiting the packed foods)	Participants: 5 families from San Francisco Bay (California). The exposure to BPA and DEHP were considerably reduced after the dietary intervention	[33]

(continued)

Table. 8.1 (continued)

Migrant/chemical	Source	Exposure estimation (Tool)	Exposure	Reference
BPA, DEHP and DiNP	Processed and packaged fast food	Determining phthalate metabolites and BPA in urinary samples	Country: US Population: participants ≥ 6 years The levels of phthalates and BPA were higher in fast-food consumers compared with non-consumers but however for BPA the difference was not statistically significant	[34]
Perfluorinated acids (PFAs) Perfluorooctanesulfonate (PFOS) Perfluorooctanoate (PFOA) Perfluorononanoate (PFNA)	Environment Food packaging	Deterministic approach using national food intake data	Country: Canada Population: participants ≥ 12 years PFAs 250 ng/day (4.0 ng/kg bw/day) PFOS 110 ng/day PFOA 70 ng/day PFNA 70 ng/day	[35]

Perfluorinated compounds Perfluorohexanesulfonate (PFHxS), PFOS, Perfluorohexanoic acid (PFHxA), and perfluorooctanoic acid (PFOA)	Food processing and food packaging	F (females), (M) (males) To do the calculations some considerations were taken into account, when the analyte was at a concentration under the LOD, three scenarios were evaluated ND = 0; ND = 1/2 LOD; ND = LOD)	Country: Spain (Tarragona) Example: marinated salmon packaged PFOS 4–9 years: (M) (ND = 0) 1.11 ng/day; (ND = 1/2LOD) 3.82 ng/day; (ND = LOD) 6.53 ng/day (F) (ND = 0) 1.10 ng/day; (ND = 1/2LOD) 3.81 ng/day; (ND = LOD) 6.51 ng/day 10–19 years: (M) (ND = 0) 0.82 ng/day; (ND = 1/2LOD) 2.83 ng/day; (ND = LOD) 4.84 ng/day (F) (ND = 0) 0.89 ng/day; (ND = 1/2LOD) 3.08 ng/day; (ND = LOD) 5.27 ng/day 20–65 years: (M) (ND = 0) 1.23 ng/day; (ND = 1/2LOD) 4.23 ng/day; (ND = LOD) 7.24 ng/day (F) (ND = 0) 1.34 ng/day; (ND = 1/2LOD) 4.62 ng/day; (ND = LOD) 7.90 ng/day > 65 years: (M) (ND = 0) 0.97 ng/day; (ND = 1/2LOD) 3.34 ng/day; (ND = LOD) 5.72 ng/day (F) (ND = 0) 0.87 ng/day; (ND = 1/2LOD) 3.02 ng/day; (ND = LOD) 5.16 ng/day	[36]

(continued)

Table. 8.1 (continued)

Migrant/chemical	Source	Exposure estimation (Tool)	Exposure	Reference
Photoinitiators (Benzophenone) (BP)	Paper and paperboard packages	Probabilistic approach (Monte Carlo simulation)	Country: Portugal Population: children 9.97E-4–3.1E-2 mg BP/day kg bw	[37]
Ag and Cu	Nanosilver and nanocopper food contact material surface coating	Probabilistic model (Monte Carlo simulation) (risk analysis software (@Risk 6.3, Palisade))	Ag: 0.00021–0.002 mg/kg bw/day Cu: 0.00038–0.0035 mg/kg bw/day	[38]
Ag	PVC nanocomposite	Probabilistic model (Monte Carlo simulation) (risk analysis software (@Risk 5.5, Palisade))	Calculated provisional ingestion limit 10 nm nanoparticles 0.084 mg Ag/kg bw/day 50 nm nanoparticles 0.421 mg Ag/kg bw/day	[39]
Total Ag Ag nanoparticles	Commercial plastic food containers	Data for a worst-case acute exposure	For both ionic Ag and Ag nanoparticles the estimated exposure is low	[40]
Al	Food and kitchenware	Nusser method (C-side software)	Country: Belgium Population: adult ≥ 15 years 0.144 mg/kg bw/day Food: 0.113 mg/kg bw/day Kitchenware: 0.031 mg/kg bw/day	[41]
Styrene	Yogurt pots	Probabilistic approach	1–35 µg/day/person	[42]

Expoxidised soybean oil (ESBO) Styrene monomer	Food packaging	CREMe® Software v.2.0.2 (CREMe Software Ltd 2006) Central Risk and Exposure Modelling e-Solution	Country: Ireland Population: children 5–12 years ESBO: (mean) 0.023 mg/kg bw/day; 0.077 mg/kg bw/day (95th percentile), 0.098 mg/kg bw/day (97.5th percentile) Styrene (maximum migration values): (mean) 0.17 µg/kg bw/day; 0.47 µg/kg bw/day (95th percentile), 0.57 g/kg bw/day (97.5th percentile)	[43]

References

[1] R. Franz, *Food Additives & Contaminants*, 2005, 22, 10, 920.

[2] E. Duffy, A.P. Hearty, M.B. Gilsenan and M.J. Gibney, *Food Additives & Contaminants*, 2006, 23, 6, 623.

[3] E. Duffy, A.P. Hearty, A. Flynn, S. McCarthy and M.J. Gibney, *Food Additives & Contaminants*, 2006, 23, 7, 715.

[4] European Commission Regulation (EU) No 10/2011 of 14th January 2011 on Plastic Materials and Articles Intended to Come into Contact with Food, *Official Journal of the European Union L*, 2011, 12, 1.

[5] L. Hammarling, H. Gustavsson, K. Svensson, S. Karlsson and A. Oskarsson, *Food Additives & Contaminants*, 1998, 15, 2, 203.

[6] L. Fantoni and C. Simoneau, *Food Additives & Contaminants*, 2003, 20, 11, 1087.

[7] *Guidance for Industry: Preparation of Premarket Submissions for Food Contact Substances: Chemistry Recommendations*, US Department of Health and Human Services, Center for Food Safety and Applied Nutrition US Food and Drug Administration, Washington, DC, USA, April 2002 and December 2007. http://www.fda.gov/Food/GuidanceRegulation/GuidanceDocumentsRegulatoryInformation/ucm081818.htm#iie

[8] E. Duffy, A.P. Hearty, S. McCarthy and M.J. Gibney, *Food Additives & Contaminants*, 2007, 24, 1, 63.

[9] P.K.T. Oldring, C. O'Mahony, J. Dixon, M. Vints, J. Mehegan, C. Dequatre and L. Castle, *Food Additives & Contaminants: Part A*, 2014, 31, 3, 444.

[10] *Towards a Harmonised Total Diet Study Approach: A Guidance Document*, European Food Safety Authority and Food and Agriculture Organization of the United Nations in *ESFA Journal*, DOI:10.2903/j.efsa.2011.2450, 2011.

[11] K. Vin, A. Papadopoulos, F. Cubadd, F. Aureli, H.I. Oktay Basegmez, M. D'Amato, S. De Coster, L. D'Evoli, M.T. López Esteban, M. Jurkovic, M. Lucarini, H. Ozer, P.M. Fernández San Juan, I. Sioen, D. Sokolic, A. Turrini and V. Sirot, *Food and Chemical Toxicology*, 2014, 73, 21.

[12] A. Ballesteros-Gómez, S. Rubio and D. Pérez-Bendito, *Journal of Chromatography A*, 2009, 1216, 449.

[13] J. Maia, J.M. Cruz, R. Sendón, J. Bustos, M.E. Cirugeda, J.J. Sánchez and P. Paseiro, *Food Research International*, 2010, 43, 1283.

[14] M.I. Santillana, E. Ruiz, M.T. Nieto, A. Rodríguez Bernaldo de Quirós, R. Sendón, M.E. Cirugeda and J.J. Sanchez, *European Food Research and Technology*, 2013, 236, 883.

[15] G. Cinelli, P. Avino, I. Notardonato, A. Centola and M.V. Russo, *Food Chemistry*, 2014, 146, 181.

[16] Y. Fan, S. Liu and Q. Xie, *Talanta*, 2014, 119, 291.

[17] M. He, C. Yang, R. Geng, X. Zhao, L. Hong, X. Piao, T. Chen, M. Quinto and D.O. Li, *Analytica Chimica Acta*, 2015, 879, 63.

[18] H.X. Sui, L. Zhang, P.G. Wu, Y. Song, L. Yong, D.J. Yang, D.G. Jiang and Z.P. Liu, *International Journal of Hygiene and Environmental Health*, 2014, 217, 695.

[19] M.F. Poças, J.C. Oliveira, J.R. Pereira and T. Hogg, *Food Additives & Contaminants: Part A*, 2010, 27, 10, 1451.

[20] M. Dickson-Spillmann, M. Siegrist, C. Keller and M. Wormuth, *Risk Analysis*, 2009, 29, 8, 1170.

[21] A. Schecter, M. Lorber, Y. Guo, Q. Wu, S. Hun Yun, K. Kannan, M. Hommel, N. Imran, L.S. Hynan, D. Cheng, J.A. Colacino and L.S. Birnbaum, *Environmental Health Perspectives*, 2013, 121, 4, 473.

[22] T. Fierens, A. Standaert, C. Cornelis, I. Sioen, S. De Henauw, H. Willems, M. Bellemans, M. De Maeyer and M. van Holderbeke, *Environment International*, 2014, 73, 117.

[23] A. Rivas, C. Monteagudo, L. Heras-González, M. Mariscal-Arcas, M.L. Lorenzo-Tova and F. Olea-Serrano, *Food and Chemical Toxicology*, 2016, 94, 25.
[24] M. Mariscal-Arcas, A. Rivas, A. Granada, C. Monteagudo, M.A. Murcia and F. Olea-Serrano, *Food and Chemical Toxicology*, 2009, 47, 506.
[25] S.R. Howe and L. Borodinsky, *Food Additives & Contaminants*, 1998, 15, 3, 370.
[26] S.R. Howe, L. Borodinsky and R.S. Lyon, *Journal of Coatings Technology*, 1998, 70, 877, 69.
[27] H.R. Dhaini and R.M. Nassif, *Environmental Monitoring Assessment*, 2014, 186, 5655.
[28] C. Liao and K. Kannan, *Environmental Science & Technology*, 2011, 45, 9372.
[29] M. Lorber, A. Schecter, O. Paepke, W. Shropshire, K. Christensen and L. Birnbaum, *Environment International*, 2015, 77, 55.
[30] A. Guart, F. Bono-Blay, A. Borell and S. Lacorte, *Food Additives & Contaminants: Part A*, 2011, 28, 5, 676.
[31] P.K.T. Oldring, L. Castle, C. O'Mahony and J. Dixon, *Food Additives & Contaminants: Part A*, 2014, 31, 3, 466.
[32] A.K. Sakhi, I.T.L. Lillegaard, S. Voorspoels, M.H. Carlsen, E.B. Løkend, A.L. Brantsæter, M. Haugen, H.M. Meltzer and C. Thomsen, *Environment International*, 2014, 73, 259.
[33] R.A. Rudel, J.M. Gray, C.L. Engel, T.W. Rawsthorne, R.E. Dodson, J.M. Ackerman, J. Rizzo, J.L. Nudelman and J. Green Brody, *Environmental Health Perspectives*, 2011, 119, 7, 914.
[34] A.R. Zota, C.A. Phillips and S.D. Mitro, *Environmental Health Perspectives*, 2016, 124, 10, 1521.
[35] S.A. Tittlemier, K. Pepper, C. Seymour, J. Moisey, R. Bronson, X.L. Cao and R.W. Dabeka, *Journal of Agricultural and Food Chemistry*, 2007, 55, 3203.
[36] I. Ericson Jogsten, G. Perelló, X. Llebaria, E. Bigas, R. Martí-Cid, A. Kärrman and J.L. Domingo, *Food and Chemical Toxicology*, 2009, 47, 1577.
[37] C. Machado, C. Fernandes, J. Pereira and M.F. Poças in the *Proceedings of the 6th International Conference on Simulation and Modelling in the Food and Bio-Industry, (FOODSIM'2010)*, Bragança, Portugal, 2010, p.241.
[38] J.C. Hannon, J.P. Kerry, M. Cruz-Romero, S. Azlin-Hasim, M. Morris and E. Cummins, *Food and Chemical Toxicology*, 2016, 95, 128.
[39] M. Cushen, J. Kerry, M. Morris, M. Cruz-Romero and E. Cummins, *Food Chemistry*, 2013, 139, 389.
[40] N. von Goetz, L. Fabricius, R. Glaus, V. Weitbrecht, D. Günther and K. Hungerbühler, *Food Additives & Contaminants: Part A*, 2013, 30, 3, 612.
[41] V. Fekete, S. Vandevijvere, F. Bolle and J. Van Loco, *Food and Chemical Toxicology*, 2013, 55, 602.
[42] O. Vitrac and J.C. Leblanc, *Food Additives & Contaminants*, 2007, 24, 2, 194.
[43] E. Duffy and M.J. Gibney, *Food Additives & Contaminants*, 2007, 24, 2, 216.

Index

https://doi.org/10.1515/9783110648065-009

Bei Fragen zur Produktsicherheit wenden Sie sich bitte an:
If you have any questions regarding product safety,
please contact:

Walter de Gruyter GmbH
Genthiner Straße 13
10785 Berlin
productsafety@degruyterbrill.com